Geographies of Digital Culture

T0179099

"Digital culture" reflects the ways in which the ubiquity and increasing use of digital devices and infrastructures is changing the arenas of human experience, creating new cultural realities. Whereas much of the existing literature on digital culture addresses the topic through a sociological, anthropological, or media theoretic lens, this book focuses on its geographic aspects.

The first section, "infrastructures and networked practices" highlights the integration of digital technologies into everyday practices in very different historical and geographical contexts—ranging from local lifeworlds, urban environments, web cartographies up to global geopolitics. The second section on "subjectivities and identities" shows how digital technology use possesses the capacity to alter the subjective, perceptive, and affective engagement with the spatial world. Finally, "politics and inequalities" investigates the social and spatial disparities concerning digital technology and its use.

This book draws attention to the deep interconnectedness of the cultural, digital, and spatial aspects of everyday practices by referring to a broad range of empirical examples taken from tourism, banking, mobility, and health. Scholars in human geography, anthropology, media and communication studies, and history will find this research indispensable reading. It addresses both young and seasoned researchers as well as advanced students in the aforementioned disciplines. The wealth of examples also makes this publication helpful in academic teaching.

Tilo Felgenhauer is lecturer at the department of social geography at the University of Jena, Germany.

Karsten Gäbler is lecturer at the department of social geography at the University of Jena, Germany.

Routledge Research in Culture, Space and Identity
Series editor: Dr. Jon Anderson
School of Planning and Geography, Cardiff University, UK

The *Routledge Research in Culture, Space and Identity Series* offers a forum for original and innovative research within cultural geography and connected fields. Titles within the series are empirically and theoretically informed and explore a range of dynamic and captivating topics. This series provides a forum for cutting edge research and new theoretical perspectives that reflect the wealth of research currently being undertaken. This series is aimed at upper-level undergraduates, research students and academics, appealing to geographers as well as the broader social sciences, arts and humanities.

For a full list of titles in this series, please visit www.routledge.com/ Routledge-Research-in-Culture-Space-and-Identity/book-series/CSI.

Memory, Place and Identity
Commemoration and Remembrance of War and Conflict
Edited by Danielle Drozdzewski, Sarah De Nardi and Emma Waterton

Surfing Spaces
Jon Anderson

Violence in Place, Cultural and Environmental Wounding
Amanda Kearney

Arts in Place
The Arts, the Urban and Social Practice
Cara Courage

Explorations in Place Attachment
Jeffrey S. Smith

Geographies of Digital Culture
Edited by Tilo Felgenhauer and Karsten Gäbler

Geographies of Digital Culture

Edited by
Tilo Felgenhauer and
Karsten Gäbler

Routledge
Taylor & Francis Group

LONDON AND NEW YORK

First published 2018 by Routledge

2 Park Square, Milton Park, Abingdon, Oxfordshire OX14 4RN
52 Vanderbilt Avenue, New York, NY 10017

Routledge is an imprint of the Taylor & Francis Group, an informa business

First issued in paperback 2019

British Library Cataloguing in Publication Data
A catalogue record for this book is available from the British Library

Library of Congress Cataloging in Publication Data
A catalog record for this book has been requested

ISBN: 978-1-138-23622-6 (hbk)
ISBN: 978-0-367-88538-0 (pbk)

Typeset in Times New Roman
by Wearset Ltd, Boldon, Tyne and Wear

Contents

Illustrations

Figure

Tables

Contributors

Pablo Abend is Scientific Coordinator of the Interdisciplinary Research School "Locating Media" at the University of Siegen (Germany). Previously he worked as a post-doctoral Researcher in the project "Modding and editor-games: participatory practices in mediatized worlds," funded by the German Research Foundation (DFG). Before that he worked as a Lecturer at the Institute for Media Studies and Theater, University of Cologne. In 2012 he received his PhD at the University of Siegen for his thesis on historical and contemporary cartographic practices. His research focuses on geomedia, localized and situated media research, game studies, qualitative methods in media studies, and science and technology studies. He is co-editor of the journal *Digital Culture & Society*.

Michael Bauder is post-doctoral Research Associate at the Institute for Environmental Social Sciences and Geography, University of Freiburg (Germany). Since 2011 he worked primarily on GIS and GIScience, tourism geography, and mobility, focusing (among others) on GPS tracking methods and their implementation in a social science perspective. Since 2014 he extended his research to the emerging fields of digital geography, volunteered geographic information (VGI), ambient geospatial information (AGI), and big data, including processes of digitization of society and scientific research.

Fabio Betioli Contel is Associate Professor at the Department of Geography at the University of São Paulo (Brazil). As a trained geographer he has worked in particular in the fields of economic geography and urban geography. He has done research on the financialization of geographical space at the universities of São Paulo and Jena (Germany).

Mike Duggan is a cultural geographer and independent researcher whom has recently been awarded a PhD from Royal Holloway University of London (UK). His research is focused on the intersections between everyday life, mapping practices and digital technologies. His most recent work "Questioning 'digital ethnography' in an era of ubiquitous computing" was published in *Geography Compass*, 15 (5) in May 2017.

Tilo Felgenhauer is Assistant Professor at the Department of Social Geography at the University of Jena (Germany). Trained in geography, history, and political science, his main research field is the linguistic and visual construction of regions and regional identities, in particular in relation to mass media. He is currently working on the question of how everyday routines (mobility, communication, and consumption), which depend on the interaction with digital interfaces, integrate both traditional and newly designed spatial symbolics.

Karsten Gäbler is post-doctoral Research Associate at the Department of Social Geography at the University of Jena (Germany). His recent work examines historical aspects of everyday digitalization processes and conceptual history approaches in geography. In particular, he investigates historical relationships between spatial concepts and metaphors, and communication and transport technologies.

Paul Montuoro is Researcher at St. Bernard's College in Melbourne (Australia), and he also works as a Researcher and Statistical Consultant in the Faculty of Education at La Trobe University in Melbourne, where he received his PhD in 2016. His research explores the role of attachment theory in education. He is also a passionate Husserlian working to understand the impact of speed of light technologies on young people's interactions with the real. His most recent journal article is titled "An investigation of the mechanism underlying teacher aggression: Testing I3 theory and the General Aggression Model" (co-authored with Tim Mainhard, *British Journal of Educational Psychology*, 2017).

Annika Richterich is Assistant Professor in Digital Culture at Maastricht (NL) University's Faculty of Arts & Social Sciences. Her research explores how social practices and media content co-evolve with digital technologies. She focuses on two main subject areas: the social and ethical implications of big data use in public health surveillance; as well as techno-social practices in hacker and maker cultures. She received her PhD in media studies from University of Siegen (Germany) in 2012. Prior to that, she studied media economics (Siegen) and sociology at the University of Auckland (NZ).

Margaret Robertson is Professor of Education in The College of Arts, Social Sciences and Commerce at La Trobe University, Melbourne (Australia). She is a member of the Steering Committee of the "International Year of Global Understanding" (IYGU) with interests in youth cultures, pedagogical change, and transforming education through innovative uses of technologies. Research interests include cross-cultural analyses of young people's views and visions for the future. She has long contributed to the curriculum and research outputs related to geographical education and youth studies.

Harald Sterly is Researcher at the University of Bonn (Germany). After completing his graduate studies of geography at the University of Cologne, he specialized in development studies at the Centre for Rural Development

(SLE) in Berlin. Currently, he acts as scientific coordinator of the junior research group "Environment, Migration, Resilience" (TransRe) at the University of Bonn. His research interests include the nexus of society, development, and technology, especially information and communication technology, as well as migration, urbanization, and translocal and rural-urban linkages. His regional foci are South and Southeast Asia as well as Western Africa.

Barney Warf is Professor of Geography at the University of Kansas (USA). His research and teaching interests lie within the broad domain of human geography. Much of his research concerns telecommunications, particularly the geographies of the internet, including the digital divide, e-government, and internet censorship. He has also studied international producer services, fiber optics, the satellite industry, offshore banking, military spending, voting technologies, the U.S. electoral college, religious diversity, and geographies of corruption. He has authored, co-authored, or co-edited eight books, three encyclopedias, 55 book chapters and more than 110 refereed journal articles. Currently, he serves as editor of *Geojournal* and *The Professional Geographer*, and co-editor of *Growth and Change*.

Roland Wenzlhuemer is Professor of Modern History at the University of Munich (Germany). He is specifically interested in the history of globalization in the "long" nineteenth century. In this context, he has recently worked on the history of the telegraph and of steamship passages. He is the author of *Connecting the Nineteenth-Century World: The Telegraph and Globalization* (Cambridge University Press, 2012), and of *Globalgeschichte schreiben: Eine Einführung in 6 Episoden* (UTB, 2017).

Acknowledgments

The idea for this book was born at a joint session of the International Geographical Union's commissions "Cultural Approach in Geography" and "Geography of the Global Information Society" in Krákow in 2014. We thank all participants, discussants, and commission members for their input and advice. We are particularly grateful to Amanda Halter, Robert Wenzl, and Dorothee Quade for their invaluable help in all of the project's stages.

1 Geographies of digital culture

An introduction

Tilo Felgenhauer and Karsten Gäbler

It is an often-heard diagnosis that the world and our everyday lives are currently undergoing major transitions (see, e.g., Floridi 2014 or Rifkin 2011). The advent of digital, especially mobile digital technologies seems to have changed nearly all domains of life, transforming the ways people share information, relate to one another, engage in political issues, act in public and private spheres, etc. Geographers in particular often stress that digital technology is closely associated with new spatialities, i.e., new perceptions of space, new spatial or corporeal practices, or new patterns of the distribution of digital devices and infrastructures. At first glance, it seems as if the so-called "Digital Turn" fundamentally altered—and continues to alter—our ways of being in the world (see Lagerkvist 2017).

From a social and cultural science perspective, however, the popular talk of a Digital Turn merits a second look. Indeed, the newly emerged digital culture and its geographies present us with previously unheard-of everyday practices. One might even say that the use of digital technology serves as a fundamental means of social integration today. In many contexts, possessing digital devices and broadband internet access means leading a "normal" life. From a geographical perspective, digital technologies derive much of their novelty from their capacity to (seemingly) deterritorialize everyday phenomena. For instance, virtual public spaces offer new forms of political expression for citizens and social movements and thus alter forms of political participation and protest (see, e.g., Kavada 2016; Lee et al. 2017; Maireder and Ausserhofer 2014). Digitally mediated communication also facilitates the simultaneous integration into dispersed communities, thus fostering multi-local lifestyles and intimate relationships from a distance (Beck and Beck-Gernsheim 2014). Digital labor additionally blurs the lines between work places and private places, producing both new forms of freedom and flexibility and new forms of exploitation (see, e.g., Brynjolfsson and McAfee 2011, 2014; Dyer-Witheford 2015; Huws 2003).

Yet, these forms of digital technology use are not in any way deterritorialized or "despatialized," but produce—as will be shown in this book—their own spatial configurations. For instance, internet or telephone network access relies upon unequally distributed material infrastructures (as the rural-urban or other broadband divides indicate); network usage is embedded in national or territorial regulatory systems (see, for instance, the different laws for regulating data traffic surveillance or the debates on geoblocking); and access to devices and infrastructures is differentiated not only spatially, but also socially and culturally (for instance, marketing lingo's "ubiquitous connectivity" does not reflect the lived experience for many people).

As we can see from these few examples, talk of a Digital Turn is ambiguous. New forms of "digital" practice intermingle with "pre-digital" forms of social organization, strategies of surveillance, regulation, exploitation, or patterns of social stratification, to name but a few.

As research in the history of technology has shown (see, e.g., Park et al. 2011 or Rid 2017), the current talk of a Digital Turn in many respects is historically shortsighted and tends to unjustifiably privilege the present. Emphasizing the novelty of real-time communication and global connectivity, for example, masks the fact that current technologies often merely perpetuate technological principles established in the mid-nineteenth century (see, e.g., Marvin 1988 and Wenzlhuemer in this volume). It is without doubt that the pervasiveness of modern digital devices—like the cellphone—and their integration into what Floridi (2014: 59ff.) has called "onlife," distinguishes them from, say, nineteenth-century telegraph use. However, "Digital Turn" diagnoses often assert fundamental transformations of society-space relationships and thus tend to lose sight of the subtler geographical and socio-cultural changes associated with digital technology use. Talk of disruptive change tends to homogenize everything before and after the (moment of) transformation—as if everything in the twenty-first century was "digital" compared to a thoroughly "analog" past. Hence, research on the Digital Turn and its geographies requires both a historical perspective and an investigation of contemporary practices constituting digital culture.

In the following sections the overall scene for the contributions presented in this book will be set. We first identify some key features of the digital, trying to grasp the conceptual core of digital technology and its use from a rather general point of view. In the second part we offer a brief overview of current geographical research on the digital in geography with particular emphasis on research in social and cultural geography. The final section outlines three fields of research on digital culture and presents the overall structure of the book.

Putting digital culture in perspective

Questions of digital technology have been around for a while in social and cultural science. Whereas the dissemination of personal computers (PCs) to a wider public since the mid-1970s and the development and common use of the World Wide Web since the early 1990s have, for instance, raised questions of organizational change (see, e.g., Berghel 1997; Morell 1988; National Academy of Engineering 1983), or questions of the growing significance of information for modern societies (see Manuel Castells' trilogy "The Information Age," and in particular Castells 1996), since at least the mid-2000s an increasing portion of literature on digital technology focuses on everyday practices and the advent of a quotidian digital culture (see, e.g., Couldry 2012: 33ff.; Horst and Miller 2013; Poster 2006). In particular, a renewed theoretical interest in the relationships of technology—conceived as software *and* hardware—and society has emerged since then (see, for instance, Berry 2014; Dolata 2013; Matthewman 2011; Miller 2011). While on the one hand acknowledging the essential role of technology for modern societies, most of these more recent positions reject the implicit and explicit technological determinisms popular in earlier accounts of the digitalization and assert theories of socio-technical *co-evolution*. One of these approaches' main questions is: How are "digital" technologies distinct from other technologies?

Surveying the numerous insightful answers offered to this question (see Miller 2011 for a particularly discerning approach), three elements seem particularly striking from a geographical point of view: (1) digitality, (2) network character, and (3) mobility.

(1) It is all but surprising that digitality and digitalization are the central terms to understand the processes commonly described as the *Digital* Turn. Despite the apparent self-evidence of these concepts, they turn out to be remarkably ambiguous upon a second view. Digitalization, for instance, serves as an umbrella term for various things (Passig and Scholz 2015). On the one hand, it refers to developments such as the rapid diffusion of large computer systems and later of PCs in public administration, hospitals, etc. since the 1970s, or to the current permeation of everyday lifeworlds with networked devices and their social, cultural, or economic effects. Such accounts of digitalization are fundamentally bound to a user's point of view; they presume technological change, but they do not address the underlying technical procedures.

On the other hand, the term "digitalization" (in a somewhat imprecise fashion) is also used to refer to the conversion of analog data into a binary numerical system, or, to code information using 1s and 0s. This notion of "digital" can be traced back to the Latin noun "digitus," meaning the finger used to count. In academic language, the process of transforming information into 1s and 0s is thus often called "digitization," as opposed to "digitalization," denoting the consequences of digitization for institutions and organizations (Brennen and Kreiss 2016). Digitization as the underlying core practice of digital culture has been discussed extensively in practical contexts, but has, of course, historically also always been a subject of mathematics or philosophy.

With regard to the transformative capacity of digitization, the translation of continuous into discrete numeric values initially seems to be of particular importance, because it comprises a fundamental reduction of complexity. This has significant consequences for technically mediated communication, because it could be said that any device able to represent two sufficiently different states may perform digital information processing. In a historical perspective then, we can observe digital communication independently of the use and availability of electric power—think, for instance, of optical telegraphic devices like the heliograph, a mirror using reflected sunlight as a communication signal, or flag telegraphy (see Maddalena and Packer 2015: 94ff.). In geographical respect, such digitally enabled detachment of information transmission from the mobility of a physical carrier facilitated long-distance communication. With the marriage of electricity and digital information encoding—best represented by the Morse code and the telegraphic system developed in the 1830s and 1840s—digital communication saw its first heyday.[1]

Another influential property of the digitization of information is that it allows for electronically enhanced computation. In fact, it seems as if many of the features currently ascribed to digital culture are rather an effect of computation than of sheer digitality. Nowadays, automated data processing—from simple algorithms through to notorious "big data" applications—shapes a big part of our technologically mediated experiences and our ways of constructing reality (see, e.g., Amoore and Piotukh 2016; Seyfert and Roberge 2016). For instance, what is being presented to us as relevant information by digital media—what news and advertisements we get to read, what search results we get, etc.—is mainly a result of algorithmically analyzed (user) data. As has been pointed out by many critical approaches (see, e.g., O'Neil 2016), algorithms in this sense (as automatically acting instances) possess agency. However, with regard to the historical dimension of digital culture it should be noted that algorithms—conceived in a general sense as pre-defined steps and instructions to solve certain tasks—are in principle independent from modern electronic technologies but can be traced back to the eighteenth century and its mechanical computation machines. In particular, Charles Babbage's "Analytical Engine," developed as early as the first half of the nineteenth century, mechanically embodies the modern-day idea of a computer as an automatic calculating machine (Burckhardt 2017: 50ff.).

Finally, processes of digit(al)ization and the respective juxtaposition of "analog" and "digital" are fuzzy in another, usually quite unexpected way. As Passig and Scholz (2015: 78f.) have pointed out, the binary opposition of "analog" and "digital" is not even convincing in a technical sense, since discrete values like 1 and 0 can only indirectly be represented in computers. Instead of clear distinctions between "power on" and "power off," computers need to operate with continuous voltage ranges and—more or less arbitrary—thresholds. What is analog and what is digital then depends on the perspective or scale the observer selects. As this book will show, there is good reason to assume that this holds true also for the distinction of an analog and a digital realm of everyday life (in particular, see Montuoro and Robertson's contribution in this volume).

(2) The second key element of digital culture to consider is network character. The integration of stationary and portable electronic devices into everyday life considerably increased with public access to global communication networks such as the internet from the early 1990s onwards (see, e.g., Kohut et al. 1995 for the U.S.-case). It can reasonably be stated that digital culture as to a large extent known today is a product of *web-enabled* (mobile) devices.[2] In today's digital practices, links to distant others are constantly created, constituting subjects as nodes in communication networks; the World Wide Web is used more or less effortlessly as a comprehensive, globally linked information system (see

Wilson et al. 2013: 48) and non-human actors are becoming more and more integrated into an internet of (intelligent) things.

Whereas it is true that the internet has massively boosted global connectivity—like the telephone and the telegraph before—and brought all kinds of computers into common use, networking as a practice, of course, is not at all bound to twentieth or twenty-first century digital (electronic) technologies. From a very general point of view, networking means nothing more than creating connections between two or more entities and setting up processes of circulation. Networks thus consist of nodes, links, and streams, constituting a topological space. In this wide sense, networking must be considered to be a fundamental, almost inevitable human practice. Yet, using the image of a network to describe interconnected entities is a rather modern affair. Giessmann (2014: 17; see also Barkhoff et al. 2004), for instance, points out that networks as *explicit* interpretative models have been used at least since eighteenth century discourses.[3] Osterhammel (2014: 710ff.) on the other hand, associates networks with the nineteenth century "transformation of the world" and in particular with the development of circulative systems in the domains of traffic and communication (e.g., regular steamship service, railway, telegraphy), commerce (e.g., stable international trade relations), and money and finance (e.g., adjusted international currency systems). According to Osterhammel, networks possess a fundamental geographical dimension and are essentially linked to globalization.

With regard to today's digital technologies, the network character seems to be particularly pertinent: Digital communication, for example, is often bidirectional or interactive—content can be exchanged by many agents, and information can be shared more or less easily. This is important to note because it multiplies the information potentially available. As Miller (2011: 15) points out, the multitude of active nodes in networks (namely, subjects producing content) increases the *diversity of choice*, thus challenging the informational monopolies represented by "old" one-way media. However, as the debates on internet surveillance or "fake news" indicate, utopian visions of a democratization through networked digital practices must be handled with great care.

Digitally mediated circulation of information between subjects, additionally, can be experienced as physical presence at a distance. In particular, the convergence of visual and audio communication channels in modern devices and the availability of broadband internet connections facilitated the seeming irrelevance of bodily locations (see Miller 2011: 31f.; for an early account on "telepresence" see Minsky 1980). Yet, as has been said above, the blending of online and offline practices in everyday life does not suspend spatial references but blurs the line between physical and virtual presence, creating its own geographies.

(3) The third key attribute of today's digital technology is mobility. With the convergence of miniaturized computing technologies and wireless internet access in smartphones and other portable devices, new practices—such as the use of "augmented reality" applications or satellite navigation—emerged. With mobile internet use even exceeding desktop use today (StatCounter 2016), "wireless" has become one of the trademarks of digital culture.

Unsurprisingly, this increased mobility calls for geographical attention. Three aspects should be highlighted here. The first refers to the material backbone of the seemingly immaterial mobile media. Despite that the internet in everyday life is often associated with virtuality, and despite wireless access reinforcing the impression of it "just being there," only the smallest part of the technical infrastructure necessary to run mobile internet access is portable. Global internet communication, for instance, relies upon an undersea cable network quite similar to the nineteenth and early twentieth century web of telegraphic cables (see Starosielski 2015), and network coverage requires a vast bulk of radio masts, base transceiver stations, etc. Likewise, the wireless internet used at home to conveniently stream music and TV series is grounded in broadband-enabled cables covering also the "last mile" to every consumer's house (see Warf's contribution in this volume). One might thus conclude, what at first sight appears to be mobile, adaptive, and individual rests on a surprisingly stationary, inflexible, and inert material basis.

It is obvious that the widely invisible material backbone of mobile digital technologies is inextricably linked to questions of "spatial justice" (see Soja 2010; Alizadeh et al. 2014). The non-ubiquity of broadband telecommunication infrastructures, for instance, points to the precarious and fragile nature of digital culture—despite all the marketing discourses promising a seamless integration of the devices into everyday life. Recent talk about subjects and practices must thus not make invisible the material dimension of mobile technologies.

Second, portable networked devices are increasingly used to gather and transmit personalized data (voluntary "self-tracking" being only one particular example; see Lupton 2016; Neff and Nafus 2016). It can be said that with the portability of personal communication devices an amalgamation of subjects and objects is put forward. Smartphones have become not only a part of people's identities—as the keepers of personal, often intimate information –, but they also serve as an extension of the body. The use of mobile devices has the capacity to change basic corporeal practices (see Byington and Schwebel 2012 on pedestrian behavior), as well as bodies themselves (see Hansraj 2014 on spine damage caused by smartphone use).

In geographical respect, the individualization made possible by personal mobile devices is ambiguous. On the one hand, individually (and often automatically) collected and accumulated geodata facilitates new forms of collective mapping and the provision of place-specific information (see below and see Abend's and Bauder's contributions in this volume). On the other hand, this "volunteered geographic information" (VGI) also presents us with ethical and legal questions of data privacy and surveillance (see, e.g., Blatt 2015).

A third aspect of digital media's mobility, finally, is the particular importance of context. Since technology use is separated here from fixed and largely expectable environments (like in the case of the landline telephone), contexts have to be made explicit in one way or another in mobile communication. As Laurier has pointed out already in 2001, cellphone calls often begin with the question "Where are you?" in order to produce (or simulate) a shared conversational context (Laurier 2001: 494). It seems to be crucial in social interactions to know about the places the respective counterpart is located in. Again, it can be observed here that digital technology use does not suspend space but produces new geographies.

Given the fundamental spatiality very briefly outlined here, it is not surprising that digital technology has become an object of great interest in geography. However, the approaches developed in the discipline's sub-branches vary considerably with regard to their conceptions of space and place, their ideas of materiality and subjectivity, and their political or emancipatory ambitions; a brief overview of which is presented in the following.

Current research on the geographies of digital technology use

The current state of geographical research on the digital contains several dimensions that are highlighted by different sub-disciplines (e.g., cultural and social geography, geoinformation science, tourism, and media geography). They are linked by theoretical debates as well as in research on particular empirical topics. Given the status of geography as a technological and, to a large extent, empirical science, it is no surprise that digital technology plays an important role in the range of methods and tools of empirical research (as did pre-digital technology before). Since the advent of geographical information systems (GIS), geography as a scientific enterprise has in itself become inherently "digital." Transcending this rather instrumental perspective, digital technology has to be transformed from being a sole means of spatial analysis toward a subject of research in its own right approached from a cultural and social science standpoint. The use and the meaning of digital technology as a part of everyday life and its particular geographies should become the focus of attention, as expressed by Dodge and Kitchin's (2005: 162) term "transductive practices," and in Leszczynski's (2015) concept of "spatial media/tion." Thus, the work of several branches of cultural geography has to be acknowledged which addresses the transformative and innovative character of digital media and technology, for example, with regard to mobility, infrastructures, consumption, perceptions of space and place, and the involved politics of participation and power. In order to give some orientation in the field of geographical research on the digital, a categorization will be applied distinguishing work on (1) the theory and the concepts to approaching the subject, (2) research on spatially distributed material infrastructures as both the outcome and condition of digital culture, (3) subjective/individual perceptions and appropriations of digital technology in the making of space and place, and (4) the social and cultural construction of space and place as both a "meta-individual" and digital phenomenon. Knowing that all of the four analytical dimensions interact or may even be in itself contested (see, e.g., Kinsley 2014), they nevertheless offer a broad overview and are able to grasp highly heterogeneous approaches when applied in a heuristic manner (which, of course, can only be illuminated with reference to a very limited number of examples here). The following sections address each of these four categories.

(1) Theoretical contributions provide abstract and general concepts to help frame and navigate the quickly changing landscape of the analysis of digital technology. Two particularly influential conceptual frameworks are shortly introduced in the following.

A well-known framework was introduced by Adams (2009, 2010) in the context of media geography following a "communicational turn" since "communications fill the world and the world fills communications" (Adams 2009: 214). Adams suggests to distinguish complementary dimensions of analysis regarding the representation of places in media, the augmentation and transformation of places by the media, the distribution of media in space (e.g., as infrastructures), and the introduction of spatial structures and concepts to media environments. His schema is especially useful when applied to different analytical dimensions of one particular empirical subject instead of assuming different, categorically separated "worlds" of digital culture contained within and limited to each of the presumed categories.

Another theoretical framework emerges as Adams' original assumption of a "communicational turn" currently turns itself into a "digital turn" (Ash et al. 2016). The term "digital geographies" is coined to address the manifold interrelations between geographical research and the digital. Ash et al. (2016) split the subject by distinguishing "geographies *through* the digital" (ibid.: 3ff.), "geographies *produced by* the digital" (ibid.: 5ff.) and "geographies *of* the digital" (ibid.: 8ff.). Geographies *through* the digital means that geographical research is based on digital information, basically since the so-called "Quantitative Revolution" of the 1960s and 1970s, the broad application of GIS afterwards, and the accessibility of "big data" resources at its current stage. This history is also critically reflected in works on origins and pathways of technological and scientific development, the emphasis on power issues, questions of equality and control, and the ongoing challenge to produce and communicate geographic knowledge digitally. Geographies *produced by* the digital investigate spatial conditions and relations as the outcome of digitalization: transformed political and economic conditions, the dependence of places, particularly cities, on digital infrastructures, the digital divide and its complication by mobile digital media, and principle questions on the origin of and the power over ubiquitous and pervasive digital information and, thus, contemporary spatial relations. Geographies *of* the digital as a third kind of geographical engagement mainly address virtual spaces (e.g., video games) as well as subjective, everyday routines reorganized with and attuned to digital devices.

The effect of reflecting on abstract theoretical efforts (like the ones introduced above) should, for instance, be an increased attention to the several potential analytical dimensions. Different aspects of acting with the digital can be highlighted and addressed more explicitly: e.g., the material and/or the symbolic; the digital as a context and/or content for action; digital technology being either a means or a subject of research. Thus, conceptual geographies of the digital may distinguish particular modes of construction. For example, semantics and imaginations contain references to and meanings of space and place or, spaces and places contain material objects and structures that put the digital into practice. Likewise, potential entanglements of the several analytical categories may be observed and applied in order to put forward the hermeneutics of geographies of digital culture.

(2) Research on the spatial distribution of material infrastructures appears as a major area implying a turn from theoretical and conceptual frameworks toward empirical work. In this sense, classic geographic epistemologies focus on spatially distributed material structures which are explored as the outcome of cultural, economic, and technological transformations. As an example from "pre-digital" times, Horvath's (1974) "machine space" was a powerful concept in naming and mapping the groundbreaking change and the seemingly limitless dominance of the car in urban space. By pointing out and mapping the vast spaces that are reserved for car use (such as maintenance and parking), the critical impacts of the dominance of a particular technology were powerfully highlighted. Partly, this effective scientific and communicative principle was retained in the shift from the pre-digital to digital technology and infrastructures. For example, Torrens (2008) offers an advanced cartography of the expansion of WiFi-signals in urban space and of spatial unevenness in the supply of the access to digital networks. Tranos (2013) offers a detailed investigation of the spatial distribution of digital infrastructures related to Europe's internet (e.g., mappings of the internet backbone networks), its standards and its topological structure, involved institutions, and the impacts for city-region development. Licoppe et al. (2008) trace and map the everyday paths of individuals in relation to their cellphone use in urban Paris.

By applying a common mode of representation—the map—to observable material formations, these kinds of research detect objects and structures in space suggesting that technology, culture, infrastructure, and the digital are contained in space. This kind of methodology promotes a realistic and objective feel of the represented objects, assuring the evidence of the material and its objective cartographic representation. According to Adams' (2010: 44ff.) distinction, empirical research on the spatial distribution of digital infrastructure could be labeled with his "media-in-space" category. Yet this branch of research may also appear as the basis for geographies "*produced* by the digital" in the sense of spatial conditions depending on the provision of digital infrastructure (Ash et al. 2016: 5ff.). Empirical studies not only follow this common cartographic logic of representation, rather, this principle is also implicit to some of the critical approaches which emphasize the power of digital infrastructures and digital media to dominate urban spaces, to guide, or even determine social and cultural processes and shifts in urban development (Thrift and French 2002; Graham 2005). Metaphors of invasion reproduce the idea of spatial expansion and dominance similar to mapping techniques.

(3) In order to augment and expand this perspective, it is helpful to focus on subjective individual perceptions and appropriations of space and place. This requires a constructivist, hermeneutic view assuming that spatial formations are basically cultural constructions. Space and place can be considered to be both the

matter and the product of individual perception and experience, cultural and linguistic convention, public discourse, and economic/political action. In this sense, space and place are not pre-given entities in which the cultural and the digital are contained. Instead, space and place depend in their existence on culturally evoked meanings and everyday action (see Sterly's qualitative fieldwork, this volume). For example, this perspective on the geographies of the digital is reflected in the discussion on subjective, individual, emotional, and affective experiences of space and place. As has been mentioned above, digital media and particularly mobile devices and graphic interfaces, offer new ways to merge spatial perception, media content, and the active involvement of the user in manifold assemblages. For example, Kwan (2007) investigated the individual perception of everyday surroundings of a Muslim woman in a post-9/11-environment in the USA by the use of digital video and GPS technology. Kwan introduced a style of cultural geography that is both scientific in its insight and political in its empathy. Digital technologies appear as a way to augment qualitative fieldwork and promote communication between researchers and their subjects. Kingsbury and Jones (2009) broadened the critical view on the digital and its geography by extending the focus to issues of power and control (see, e.g., Crampton 2008) in emphasizing the affective and emotional involvement digital (geo)media may offer. The joy of being entertained, aesthetically pleased, and playfully engaged in a world of digital geoinformation has to be acknowledged alongside structural questions on technological transformation, matters of access, power, and control. Graham et al. (2013) show—with reference to empirical vignettes—that the digitalization of urban environments does not only depend on material infrastructures, hardware, and potential access but on individual abilities, knowledge, recognized affordances, and, especially, the level of integration in new social networks. Abend and Harvey (2017; see also Abend's contribution in this volume) highlight the character of affective and immersive digital geomedia (especially geobrowsers) encouraging subjective engagement (e.g., user-specific practices of appropriation and configuration of interfaces and interaction).[4]

(4) With regard to the "meta-individual," (i.e., social and cultural dimension of the geographies of digital culture) several bodies of work become relevant such as research on social and spatial inequalities in adopting and distributing digital technology, issues of surveillance and power, and the general presence of geographical imaginations in (digital) media which form the world view on and within particular cultural public contexts. More specifically, the relevance of location-based information services for social networks and local communities, the advent and the rising popularity of volunteered geographic information (related to terms such as "neogeography" or the "geoweb"), and the role which crowdsourced geoinformation plays for political participation call for in-depth geographical research and critique. They all, to an increasing degree, depend on the logics and the functioning of digital code and algorithms (see Richterich's contribution in this volume on health issues and crowdsourced data).

The impact on everyday routines is especially striking in the case of location-based services and location-based social networks. Smartphone applications, which are sensitive to the user's geospatial location, filter and arrange bodies of information according to algorithmic procedures, a presumed relevance for the user and expected or predicted individual preferences (here, computation or algorithmization and mobility come together). Furthermore, in transcending subjective perception (see above) and "ego-centric" worldviews, these practices also promote the coordination of action and, thus, help to share a sense of place in relation to social networks and communities (Evans 2015; Farman 2012: 56ff.). This prospect is also an incentive to participate in the production and representation of volunteered geographic information, or VGI (Goodchild 2007; Poorthuis and Zook 2014). Projects such as OpenStreetMap (OSM) blur the boundaries between professional cartography and user engagement, boost communal activity, and help to turn passive consumers into enabled and knowing *prosumers* or *produsers*.[5] While appreciating the obvious social and cultural merits of VGI, cultural geography and, especially, critical GIS also investigate the more ambivalent outcomes and the inequalities in participation, empowerment, and the active application of user-generated content. Harvey (2013) stresses the important difference between crowdharvested geodata that is automatically sent from cellphones to centralized and opaque servers and voluntarily provided geoinformation, which is contributed to platforms such as OSM. Issues of participation are intensively discussed given the social and cultural unevenness of engagement, still highly specialized technological knowledge and, thus, the appearance of particular hegemonic elites which tend to govern these new contexts of the geoweb (Elwood 2010; Perkins 2014). Haklay (2013) limits expectations on a broad democratizing effect of the spread of volunteered, crowdsourced geoinformation and participatory mapping. Stephens (2013) points to significant inequalities in participation with regard to gender and, thus, conventionally mainstreamed geographies as its outcome. The polyvalent character of digital information is also subject to Parks' (2009) work on potentials and pitfalls of the geovisualization of political conflicts in Google Earth. The relationship between culturally established geographical imaginations, digital media, political information, the enlightenment of the public, and active involvement becomes visible in the mapping of crises.

What makes this broad and heterogeneous body of work interesting is the sensitivity to ambivalences and many open horizons for both conceptual and empirical work. For example, the question of persistence versus a significantly changing shape of the manifold geographies of digital divides remains a vital field of study: with regard to access and provision of digital infrastructures, issues of freedom in its use, differences stemming from cultural contexts and preferences, gender asymmetries, and the social and spatial unevenness in the distribution of education, knowledge, and skills (see the contributions of Warf and Contel in this volume). Likewise, basic socio-economic relations have to be re-examined under new conditions of digitalization. Common notions of the division of labor and corresponding geographies have to be revised. For example, the role of institutions, experts, and expert knowledge has to be reevaluated as well as the role of lay people and informal communities in the development, promotion, and distribution of digital technology. Also, the ambivalent impacts of mobile digital media on everyday geographies leave us with a feeling of the fading of the presence, immediacy, and relevance of our physical surrounding. At the same time, augmented reality applications offer new possibilities of becoming involved with our environment, providing literally location-based content and, thus, a buzzing world of local networks and communities. It will be especially interesting to see how isolating "filter bubbles" (Pariser 2011) interact with the dynamic development toward a constant touch with the spatial and communal surrounding. Finally, the matter of "glocalization" (Robertson 1995) will also be an ongoing issue moving in between and across scenarios of the annihilation of space, a flat world, and the construction of complex geographical imaginations as a part of media content. We can easily see how an overwhelming provision of information via digital channels forms our worldviews and our everyday routines both in a political and a geographical sense.

Organization of the book

Investigating the spatial dimension of digital culture, one might conclude then, is a genuine task of geography as a discipline. From a geographical point of view, questions of space are not back on the agenda, as is sometimes claimed in the notorious "spatial turn" debates, but they have never been absent. However, as the above-mentioned research in social and cultural geography has shown, what counts as "spatial"—and thus qualifies as a proper geographical subject matter—has changed considerably in the recent past. Going beyond a traditional, merely instrumental perspective in Digital Turn research by dealing with digita*l*ization rather than digitization requires drawing on the expertise of the social and cultural sciences. Somewhat paradoxically, one might thus say that reflecting *geographically* on the Digital Turn must necessarily be a multidisciplinary enterprise.

Consequently, this book presents investigations into the geographies of digital culture from various disciplines and scientific fields (e.g., geography, history, media studies). The material is divided into three broader subject areas that form the overall structure of the book (see also each section's introduction). As with many divisions, the fields are not mutually exclusive; rather, they put different analytical emphases on the same lifeworld phenomena.

The *first* field is "Infrastructures and networked practices" (Chapters 2, 3, and 4). The contributions presented here address the integration of digital technologies into everyday routines and the emergence of novel, digitally mediated practices in different historical and geographical contexts as well as in different spheres of everyday life. The chapters exemplify the various ways in which access to digital devices or infrastructures shapes perceptions and representations of space and place, and produces new forms of geodata. Eventually, the chapters point to the consequences of (not) being a node in digital communication networks, or in other words, of being (dis-)connected.

The *second* subject area is "Subjectivities and identities" (Chapters 5, 6, and 7). Digital technology use possesses the capacity to alter the subjective, perceptive, and affective engagement with the world. This begs the question of positioning ourselves—in both a metaphorical and a literal sense –, and also raises questions of the tensions between digital and non-digital modes of getting in touch with the world. Hence, the chapters presented in this section explore the ways in which digital technologies (and in particular geomedia) become an intimate part of mundane practices, paving the way for new forms of subjectivation and identity formation. Furthermore, the section also exemplifies some of the methodological challenges connected to the digitally mediated attachment to (and sometimes detachment from) places, putting particular emphasis on ethnographic fieldwork.

The *third* section "Politics and inequalities" (Chapters 8, 9, and 10) investigates the disparities connected to digital technology and its use. Expanding common notions of the digital divide in geographical research, the digital world's "gaps" are presented here as inherently plural, entangling spatial with social and cultural categories. The three chapters in this section point out how digital technology use produces and consolidates social stratifications (e.g., through access to financial markets or through infrastructure availability) and thus serves as a means for inclusion and exclusion. They also show the ways in which digital technologies are used to govern the social (e.g., in health crisis situations) and thus the chapters ultimately allude to the ethical and political dimension of digital culture.

Notes

1 Together with the co-evolving railway system it contributed to the nineteenth century "conquering of space."
2 The iPhone, introduced in 2007, being the trailblazer. Burgess (2012) even describes the rollout of the iPhone as a "Cultural Moment."
3 On the level of *implicit* knowledge and practices, however, Giessmann (2014: 17) asserts that networks play a major role since ancient times.
4 In a broader sociological sense, individualized media use may also correspond with a shift from "little boxes" of spatial containment toward a potentially global "networked individualism" (Wellman 2002).
5 At the same time, however, it must be mentioned that a significant participation inequality in OSM can be observed. Neis and Zipf (2012), for instance, point toward the fact that only about 5 percent of all registered users actively contribute to OSM—the vast majority is passively "watching."

References

Abend, P. and Harvey, F. (2017) "Maps as geomedial action spaces: considering the shift from logocentric to egocentric engagements," *GeoJournal*, 82 (1): 171–183. http://dx.doi.org/10.1007/s10708-015-9673-z.
Adams, P. (2009) *Geographies of media and communication*, Malden: Wiley & Blackwell.
Adams, P. (2010) "A taxonomy for communication geography," *Progress in Human Geography*, 35 (1): 37–57.

Alizadeh, T., Sipe, N., and Dodson, J. (2014) "Spatial planning and high-speed broadband: Australia's national broadband network and metropolitan planning," *International Planning Studies*, 19 (3–4): 359–378.

Amoore, L. and Piotukh, V. (eds) (2016) *Algorithmic life: calculative devices in the age of big data*, New York: Routledge.

Ash, J., Kitchin, R., and Leszczynski, A. (2016) "Digital turn, digital geographies?," *Progress in Human Geography*, Online first. https://doi.org/10.1177/0309132516664800.

Barkhoff, J., Böhme, H., and Riou, J. (eds) (2004) *Netzwerke: Eine Kulturtechnik der Moderne*, Köln: Böhlau.

Beck, U. and Beck-Gernsheim, E. (2014) *Distant love*, Cambridge: Polity Press.

Berghel, H. (1997) "Email: the good, the bad, and the ugly," *Communications of the ACM*, 40 (4): 11–15.

Berry, D. M. (2014) *Critical theory and the digital*, New York: Bloomsbury.

Blatt, A. J. (2015) "The benefits and risks of volunteered geographic information," *Journal of Map and Geography Libraries*, 11 (1): 99–104. http://dx.doi.org/10.1080/15420353.2015.1009609.

Brennen, J. S. and Kreiss, D. (2016) "Digitalization" in Bruhn Jensen, K. and Craig, R. T. (eds) *The international encyclopedia of communication theory and philosophy, Volume I: A–D*, Chichester: Wiley Blackwell, 556–566.

Brynjolfsson, E. and McAfee, A. (2011) *Race against the machine: how the digital revolution is accelerating innovation, driving productivity, and irreversibly transforming employment and the economy*, Lexington: Digital Frontier Press.

Brynjolfsson, E. and McAfee, A. (2014) *The second machine age: work, progress, and prosperity in a time of brilliant technologies*, New York: W. W. Norton.

Burckhardt, M. (2017) "Eine kleine Geschichte der Digitalisierung," *Merkur*, 71 (816): 47–61.

Burgess, J. (2012) "The iPhone moment, the Apple brand and the creative consumer: From 'hackability and usability' to cultural generativity" in Hjorth, L., Richardson, I., and Burgess, J. (eds) *Studying mobile media: cultural technologies, mobile communication, and the iPhone*, New York: Routledge, pp. 28–42.

Byington, K. W. and Schwebel, D. C. (2012) "Effects of mobile internet use on college student pedestrian injury risk," *Accident Analysis and Prevention*, 51: 78–83. https://doi.org/10.1016/j.aap. 2012.11.001.

Castells, M. (1996) *The rise of the network society, the information age: economy, society, and culture Vol. 1*, Malden: Blackwell.

Couldry, N. (2012) *Media, society, world: social theory and digital media practice*, Cambridge: Polity Press.

Crampton, J. (2008) "The role of geosurveillance and security in the politics of fear" in Sui, D. (ed.) *Geospatial technologies and homeland security*, Dordrecht: Springer, 283–300.

Dodge, M. and Kitchin, R. (2005) "Code and the transduction of space," *Annals of the Association of American Geographers*, 95 (1): 162–180.

Dolata, U. (2013) *The transformative capacity of new technologies: a theory of sociotechnical change*, London: Routledge.

Dyer-Witheford, N. (2015) *Cyber-proletariat: global labour in the digital vortex*, Toronto: Between the Lines.

Elwood, S. (2010) "Geographic information science: emerging research on the societal implications of the geospatial web," *Progress in Human Geography*, 34 (3): 349–357.

Evans, L. (2015) *Locative social media: place in the digital age*, Basingstoke: Palgrave Macmillan.

Farman, J. (2012) *Mobile interface theory: embodied space and locative media*, London: Routledge.

Floridi, L. (2014) *The fourth revolution: how the infosphere is reshaping human reality*, Oxford: Oxford University Press.

Giessmann, S. (2014) *Die Verbundenheit der Dinge: Eine Kulturgeschichte der Netze und Netzwerke*, Berlin: Kadmos.

Goodchild, M. (2007) "Citizens as sensors: the world of volunteered geography," *GeoJournal*, 69 (4): 211–221.

Graham, M., Zook, M., and Boulton, A. (2013) "Augmented reality in urban places: contested content and the duplicity of code," *Transactions of the Institute of British Geographers*, 38 (3): 464–479.

Graham, S. (2005) "Software-sorted geographies," *Progress in Human Geography*, 29 (5): 562–580.

Haklay, M. (2013) "Neogeography and the delusion of democratisation," *Environment and Planning A*, 45 (1): 55–69.

Hansraj, K. K. (2014) "Assessment of stresses in the cervical spine caused by posture and position of the head," *Surgical Technology International*, 25: 277–279.

Harvey, F. (2013) "To volunteer or to contribute locational information? Towards truth in labeling for crowdsourced geographic information" in Sui, D., Elwood, S., and Goodchild, M. F. (eds) *Crowdsourcing geographic knowledge: volunteered geographic information (VGI) in theory and practice*, Dordrecht: Springer, 31–42.

Horst, H. A. and Miller, D. (eds) (2013) *Digital anthropology*, London: Bloomsbury.

Horvath, R. (1974) "Machine space," *The Geographical Review*, 64 (2): 166–187.

Huws, U. (2003) *Cybertariat: virtual work in a real world*, New York: Monthly Review Press.

Kavada, A. (2016) "Social movements and political agency in the digital age: a communication approach," *Media and Communication*, 4 (4): 8–12.

Kingsbury, P. and Jones III, J. P. (2009) "Walter Benjamin's Dionysian adventures on Google Earth," *Geoforum*, 40 (4): 502–513.

Kinsley, S. (2014) "The matter of 'virtual' geographies," *Progress in Human Geography*, 38 (3): 364–384.

Kohut, A., Bowman, C., and Petrella, M. (1995) "Technology in the American household: Americans going online: explosive growth, uncertain destinations," available online at: www.people-press.org/1995/10/16/americans-going-online-explosive-growth-uncertain-destinations/ (accessed July 13, 2017).

Kwan, M. (2007) "Affecting geospatial technologies: toward a feminist politics of emotion," *The Professional Geographer*, 59 (1): 22–34.

Lagerkvist, A. (2017) "Existential media: toward a theorization of digital thrownness," *New Media & Society*, 19 (1): 96–110.

Laurier, E. (2001) "Why people say where they are during mobile phone calls," *Environment and Planning D: Society and Space*, 19 (4): 485–504.

Leczszynski, A. (2015) "Spatial media/tion," *Progress in Human Geography*, 39 (6): 729–751.

Lee, F. L. F., Chen, H.-T. and Chan, M. (2017) "Social media use and university students' participation in a large-scale protest campaign: the case of Hong Kong's umbrella movement," *Telematics and Informatics*, 34 (2): 457–469.

Licoppe, C., Diminescu, D., Smoreda, Z., and Ziemlicki, C. (2008) "Using mobile phone geolocalisation for 'socio-geographical' analysis of co-ordination, urban mobilities, and social integration patterns," *Tijdschrift voor Economische en Sociale Geografie*, 99 (5): 584–601.

Lupton, D. (2016) *The quantified self*, Cambridge: Polity Press.

Maddalena, K. and. Packer, J. (2015) "The digital body: telegraphy as discourse network," *Theory, Culture & Society*, 32 (1): 93–117.

Maireder, A. and Ausserhofer, J. (2014) "Political discourses on Twitter: networking topics, objects, and people" in Weller, K., Bruns, A., Burgess, J., Mahrt, M., and Puschmann, C. (eds) *Twitter and society*, New York: Peter Lang, 305–318.

Marvin, C. (1988) *When old technologies were new: thinking about electric communication in the late nineteenth century*, Oxford: Oxford University Press.

Matthewman, S. (2011) *Technology and social theory*, New York: Palgrave Macmillan.

Miller, V. (2011) *Understanding digital culture*, London: Sage.

Minsky, M. (1980) "Telepresence," *Omni*, 2 (9): 45–52.

Morell, J. A. (1988) "The organizational consequences of office automation: refining measurement techniques," *ACM SIGMIS Database*, 19 (3–4): 16–23.

National Academy of Engineering (1983) *The long-term impact of technology on employment and unemployment*, Washington, DC: National Academy Press.

Neff, G. and Nafus, D. (2016) *Self-tracking*, Cambridge, MA: MIT Press.

Neis, P. and Zipf, A. (2012) "Analyzing the contributor activity of a volunteered geographic information project: the case of OpenStreetMap," *ISPRS International Journal of Geo-Information*, 1 (2): 146–165.

O'Neil, C. (2016) *Weapons of math destruction: how big data increases inequality and threatens democracy*, New York: Crown.

Osterhammel, J. (2014) *The transformation of the world: a global history of the nineteenth century*, Princeton: Princeton University Press.

Pariser, E. (2011) *The filter bubble: what the internet is hiding from you*, New York: Penguin.

Park, D. W., Jankowski, N. W., and Jones, S. (eds) (2011) *The long history of new media: technology, historiography, and contextualizing Newness*, New York: Peter Lang.

Parks, L. (2009) "Digging into Google Earth: an analysis of 'crisis in Darfur'," *Geoforum*, 40 (4): 535–545.

Passig, K. and Scholz, A. (2015) "Schlamm und Brei und Bits: warum es die Digitalisierung nicht gibt," *Merkur*, 69 (798): 75–81.

Perkins, C. (2014) "Plotting practices and politics: (im)mutable narratives in OpenStreetMap," *Transactions of the Institute of British Geographers*, 39 (2): 304–317.

Poorthuis, A. and Zook, M. (2014) "Spaces of volunteered geographic information" in Adams, P. and Craine, J. (eds) *Ashgate research companion media geography*, London: Routledge, 311–328.

Poster, M. (2006) *Information please: culture and politics in the age of digital machines*, Durham, NC: Duke University Press.

Rid, T. (2017) *Rise of the machines: the lost history of cybernetics*, Melbourne: Scribe.

Rifkin, J. (2011) *The third industrial revolution: how lateral power is transforming energy, the economy, and the world*, New York: Palgrave Macmillan.

Robertson, R. (1995) "Glocalization: time-space and homogeneity-heterogeneity" in Featherstone, M., Lash, S., and Robertson, R. (eds) *Global modernities*, London: Sage, 25–44.

Seyfert, R. and Roberge, J. (eds) (2016) *Algorithmic cultures: essays on meaning, performance and new technologies*, London: Routledge.

Soja, E. W. (2010) *Seeking spatial justice*, Minneapolis: University of Minnesota Press.

Starosielski, N. (2015) *The undersea network*, Durham, NC: Duke University Press.

StatCounter (2016) "Internet usage worldwide October 2009–October 2016," available online at: http://gs.statcounter.com/press/mobile-and-tablet-internet-usage-exceeds-desktop-for-first-time-worldwide (accessed July 13, 2017).

Stephens, M. (2013) "Gender and the GeoWeb: divisions in the production of user-generated cartographic information," *GeoJournal*, 78 (6): 981–996. https://doi.org/10.1007/s10708-013-9492-z.

Thrift, N. and French, S. (2002) "The automatic production of space," *Transactions of the Institute of British Geographers*, 27 (3): 309–335.

Torrens, P. M. (2008) "Wi-Fi geographies," *Annals of the Association of American Geographers*, 98 (1): 59–84.

Tranos, E. (2013) *The geography of the internet: cities, regions, and internet infrastructure in Europe*, Cheltenham: Edward Elgar.

Wellman, B. (2002) "Little boxes, glocalization, and networked individualism" in Tanabe, M. van den Besselaar, P., and Ishida, T. (eds) *Digital cities II: computational and sociological approaches*, Berlin: Springer, 10–25.

Wilson, M. I., Kellerman, A., and Corey, K. E. (2013) *Global information society: technology, knowledge, and mobility*, Lanham: Rowman & Littlefield.

Part I

Infrastructures and networked practices

Each of the three studies in this section deals with digitally enabled ways to act across distance. Though highly different in their approach, they all tell stories about the cultural and social embeddedness of digital technology in the sense that technology has meanings attached to it and is often interpreted and adapted in creative and surprising ways. The studies are guided by a set of questions such as: In which way has the digital been integrated in everyday practices? What does it mean to be connected or disconnected? What are the geographies of being connected? The examples range from world (geo) politics, to getting in touch with the family via cellphone to strolling through a city as a tourist. Although these practices stem from very heterogeneous contexts and operate at different scales they all inform our understanding of the geographies of digital culture.

Roland Wenzlhuemer's contribution "Telegraphy and global space" pursues an archaeologic approach. In his investigation on the origins of telegraphy in the nineteenth century he builds up historic sensitivity toward the affordances and the resistance of technological artifacts transcending the mere consideration of general "impacts" of technological innovation. Telegraphic signals as the archetype of code became a game changer in military technology and global geopolitics. Vignettes from this history show the polyvalent character of the interplay of expectations, strategies, and the often unforeseen role of technology (especially in war and world politics), as well as the mastering of space and distance. Formerly marginal places step into the spotlight of the emerging global communication network and, thus, provide the stage for some surreal footnotes of world history.

Ambient geospatial information (AGI)—geoinformation that is automatically generated from posts of photographs and texts on online social networks—is the subject of Michael Bauder's text "Using social media as a big data source for research." The digital appears as a trace in space and, thus, a means for geographical investigation. He discusses both the limits and the potential of this kind of digital geoinformation with reference to tourism studies. AGI could open up new horizons in empirical investigation and mapping while at the same time it may help us to think critically about any hasty conclusion derived from big data resources. According to Bauder, its creative integration into ethnographic research designs and its combination with other methods could be especially promising.

The developing world has often been subject to the discussion and investigation of a real *and* imagined global divide. Jumping from this very broad and general view to the real world of everyday life in Bangladesh, Harald Sterly's piece on "Regionalization revisited" finds people and practices which appropriate the cellphone as an exciting means to network, connect, integrate, and stabilize communities as well as romantic relationships across space. Especially for traditional rural communities, which have to adapt to a globalized world of industrial production, mobility constraints foster the building up of new ties at the same time as they lead to the creation of new customs and new places of communication in relation to the manifold challenges of getting along and keeping in touch.

2 Telegraphy and global space

Roland Wenzlhuemer

Space in global history

The ongoing process of digitalization has deep socio-cultural implications. It gives rise to a whole bundle of new norms, practices, or values that often are referred to as digital culture. One ubiquitous element of such a digital culture can be found in the changing significance of geographical space in cultural practices. In short, geographical distance seems to play less and less of a role in a thoroughly digitized world. Space is on the retreat; its limitations have been overcome—or at least this is what often-heard popular diagnoses about our digital age seem to suggest. This reminds the historian of similar discussion about the purported annihilation of space that accompanied the emergence of the telegraph in the nineteenth century. Around the middle of the century, precisely such claims were made regarding the telegraph's capacity to overcome the spatial limitations of human communication. Accordingly, it is necessary to take a good look back at nineteenth-century processes of globalization and proto-digitalization in order to add some more historical substance to the current debate about space and digitalization. In order to do so, the following considerations draw on a relatively young subfield of historical inquiry. Global history examines how historical actors create and sustain global connections and how these in turn impact on the thoughts, feelings, and actions of the actors. In this context, global history has a pronounced interest in the creation and transformation of global spaces.

"Global" is a spatial concept. It takes the globe as its frame of reference. The term can be understood in this sense as meaning that a phenomenon is always global when it relates to the entire globe. If one were to adhere to such an understanding of the term, it would have direct consequences for global history. Such a history would have to study its subjects in as many contexts as possible and demonstrate that they truly have worldwide significance. From this perspective global space would constitute the framework for an investigation that has to be filled out as fully as possible. In actuality, global history is based on another, relational understanding of "global." The term describes phenomena that are directly connected with one another over great distances and for which, to add a further qualification, this bridged distance is a defining feature. It is these transregional connections—the basic elements for studying global history—that create global spaces.

In postmodern theory, the idea that space is not a firm, objective, or objectifiable category has long since been established. Rather, it is assumed that space is socially and culturally constituted and constantly renegotiated (Harvey 1973, 1990; Lefebvre 1991; Massey 1994, 2005; Soja 1996). Likewise, the much-cited spatial turn in the humanities and cultural sciences is based on such a dynamic concept of space (Döring and Thielmann 2008; Middell and Naumann 2010). Space is produced by social action, through the ascription of meaning and position and the constant affirmation and reproduction of the same. According to this understanding, geographical space is just one possible space among many. Like every other kind of space, it comprises the sum of all the relations between its objects, with geographical distance constituting the crucial relationship here. Regarding our notions of space, geographical space has a privileged position because it has a dominant place in human perception. Our senses position the information they present to us in geographical space. With that, it is decisive for our images of space. Yet, a geographical localization like this of our sensory input is not imperative, as is shown for instance by the progressive use of digital communications media, (Löw 2001: 93f.) which work according to completely different spatial principles. Already, this simple example demonstrates that although our notions of space—shaped by our senses—derive from geographical space; the actual network of relations is produced socially and culturally and is not congruent with geographical space.

The British geographer Doreen Massey has summed up such a notion of space in her programmatic book, *For space*, by presenting a number of proposals for how we should conceive it. First of all, space is, in her opinion, a product of mutual relationships. Space does not exist prior to identities/entities (in line with Massey's concept of subjects and objects in space) and their relationships. Instead, these identities/entities, the relationships between them, and their spatiality are all co-constitutive (Massey 2005: 10). As such, they cannot be thought of apart from one another. Following this, Massey concludes that if space is the product of mutual relationships, it must inevitably be based on plurality (ibid.: 10f.). However, while Massey chiefly talks of a space that can permit the widest diversity of relationships, it is beneficial for our analysis of global history to assume a plurality of spaces right from the outset. The historian Karl Schlögel says that the "pluralisation of space ... has something bewildering about it," but that it ultimately restores to us an "inkling of the complexity ... that is the world." "One could sum up," says Schlögel, "by saying there are as many spaces as there are subject areas, topics, media, and historical actors" (Schlögel 2003: 69; translation by Malcolm Green). Likewise, viewing this approach in the abstract and linking it with the interest in connections evinced by global history, it may be assumed that every type of connection has its own space (Wenzlhuemer 2012: 42). This kind of understanding automatically results in a great dynamism in the individual spaces and thus in space as such. Massey pins this down in her third proposal for understanding space when she says that space undergoes constant change. "It is never finished; never closed. Perhaps we could imagine space as a simultaneity of stories-so-far" (Massey 2005: 9).

The major aspects of such a conception of space may be summarized as follows: space is not a given but is socially manufactured. It arises from the connections between its subjects and objects. As such, there is not just one space but as many spaces as there are kinds of connections. Since connections change, spaces also constantly change. Finally, spaces share their subjects and objects and thus are linked with one another. For instance, historical actors are simultaneously involved in a large number of highly differing spaces.

The plurality of spaces, their dynamics, and the resulting shifts can be shown quite vividly by the example of telegraphy. This technology played a central role in the great intensification of globalization witnessed in the nineteenth century. It has been compared in its impact and functioning with, among other things, the internet (Standage 1998). Closer inspection reveals this comparison to be rather inaccurate (Wenzlhuemer 2012: 7ff.), but it is nevertheless true when it comes to the contemporary perception of the two technologies. Both of them stand for a massive acceleration of global communications, for what is referred to as the "shrinking of the world" or even the "annihilation of space and time." However, the telegraph no more reduced the size of the world or indeed conquered space and/or time than the internet has done today. It simply sped up a few select fields of communication. The telegraph created new possibilities for connections and with that new global spaces that corresponded through their actors with other spaces. It is precisely in this shift in the spatial configuration that the new, important qualities of telegraphy become clear, as will be illustrated by the following examples.

The telegraph and the purported annihilation of space

The idea of harnessing electricity to transfer information was already toyed with by inventors in the mid-eighteenth century. In 1753, an author who chose to remain anonymous apart from the initials C. M., with which he signed his article in *The Scots Magazine*, described a very concrete method for the rapid transmission of information. The apparatus was based on the scientific knowledge of the time: electricity could be produced with the assistance of an electrostatic generator and it was known that electrostatic energy could attract very light objects, such as small pieces of paper. In keeping with this, C. M. suggested setting up a separate electric circuit for each of the letters of the alphabet and assigning a little docket with a certain letter to each one. When electricity flowed from the generator along one of the circuits, the docket with the corresponding letter would be attracted electrostatically at the receiver's end (Huurdeman 2003: 48). It is not too surprising given the circuitous and unreliable means involved in this method that it failed to establish itself. However, the idea came at a time when the knowledge of electricity and its principles began to steadily expand. Scientists, such as Luigi Galvani, Alessandro Volta, or somewhat later Hans Christian Ørsted and Michael Faraday, made a considerable contribution to the understanding of electricity. Building on their insights, the early nineteenth century soon saw a new generation of scientists and engineers who devised and tried out new methods for electric data transmission. Some, such as Samuel Thomas von Soemmering, Francis Ronalds, Carl Friedrich Gauss, Wilhelm Eduard Weber, Carl August von Steinheil, and Paul Schilling von Canstatt, proved highly successful in their work. However, the technology mostly involved in these devices lacked the necessary maturity to reliably work outside of the workshop or laboratory.

Others, however, picked up on these fledgling attempts. In 1837, two functional telegraph prototypes were presented independently of one another in Great Britain and the United States. The respective designs differed strongly but both employed an electromagnet in order to make the electric signals visible. Samuel Morse presented his device in a New York seminary and subsequently refined it with Alfred Vail. However, it took Morse many years before he could convince Congress to fund an initial telegraph line between Washington and Baltimore. In London, Charles Wheatstone and William Fothergill Cooke presented their two-needle telegraph along a section of the London and Birmingham Railway. The technology was able to help in coordinating trains, but it also took a long time before railway companies were convinced of the practicality of the telegraph and they received a licence to construct a telegraph line.

In 1842, Congress finally agreed to finance a Morse line between Washington and Baltimore, and two years later, the telegraph line was opened. Almost at the same time, Cooke was at last able to persuade the Great Western Railway Company to let him erect a line at his own cost along the stretch between London Paddington and Slough. Both undertakings swiftly proved to be successful and profitable, and the railway companies gradually realized how telegraphy could help coordinate their trains and began to erect telegraph lines along their tracks. This led to a veritable telegraph mania in both the USA (Hochfelder 2012; John 2010) and Great Britain (Barton 2010; Kieve 1973; Perry 1997; Roberts 2007), which was to last through the 1840s and 1850s. During this period, national telegraph networks were set up in both countries and shortly after in broad areas of the European continent. Already at the beginning of these developments, attempts were made to link up the various national networks. In 1851, a cable was laid across the English Channel to create a telegraphic link between France and Great Britain. This was followed by additional telegraph cables between Great Britain and the continent. From the mid-1850s onward, there were repeated attempts to lay a transatlantic cable between Ireland and Newfoundland and to at last connect Great Britain and the United States. Apart from a short-lived link in 1858, all of these attempts failed. Regardless of these failed attempts, a link was established between Europe and India in 1865, and a year later a functioning connection was at last laid across the Atlantic (Holtorf 2013; Müller-Pohl 2010). From this moment on we can talk of a global telegraph network, which over the following decades was to spread across the entire world (Hugill 1999; Wenzlhuemer 2012; Winseck and Pike 2007).

The reason behind this rapid spread in telegraphy, first on the national and then on the global level, was above all the enormous practical benefits that telegraphy bestowed on various realms. It is not surprising that businessmen were among the first to show a lasting enthusiasm for telegraphy. Additionally, the army, the press, and the British colonial administration were able to derive many benefits from the new technology. The crucial push, however, came, after some initial hesitation, from the railway companies, at least in Great Britain. After Cooke managed to grasp a licence from the Great Western Railway to put his line to Slough, it was soon apparent how useful the telegraph was, above all for coordinating trains on sections of single track railway. Thanks to the new communications medium, trains could now be coordinated safely and efficiently on such stretches in the two directions (Kieve 1973: 33).

This symbiosis with the railways, whose rights of way eased in turn the construction of the telegraph lines, became the initial impulse for the rapid expansion of telegraphy. It points directly to the innovation that this new technology brought with it—the dematerialization of the information flow. With the telegraph, data transmission by electrical means had come to technical maturity. Information could now be encoded into electrical impulses using various coding systems. These impulses were then passed along an electrical conductor and finally decoded on their arrival at the other end. The telegraph made communication and transport independent of one another (Carey 1983). An electrical transmission was free of many of the restrictions that were imposed on material transportation, even over very long distances. Thus, the length of the communication route had an almost negligible effect on the time a message took. Once a functional cable had been laid, the distance that was to be negotiated had scarcely any more relevance. This resulted in a considerable increase in the speed of information exchange. Whereas a letter sent by steamer required up to ten days to cross the Atlantic, with the telegraph an answer from overseas could be received under favorable conditions within a few hours. Still more important than this reduction in absolute communication time, was the relative acceleration facilitated by the telegraph. Information could now be sent by a medium that was faster than the railway or—for intercontinental correspondence—a steamer. The telegraph placed a technology in people's hands that allowed railways and steamers to be efficiently controlled and coordinated. It was this ability and the resulting interaction with established modes of transport and communications that was the unique feature of telegraphy.

The dematerialization of information flow by telegraphic means was not only of great importance for speeding up communications (not only but most importantly concerning transport), but also it had an effect on the associated means of communication. The telegraph conveyed information as a series of electrical impulses. At times, the fact that data transmission involved electricity either flowing or not flowing along the cable has been interpreted as an early form of binary code. In fact, both Morse telegraphy and the Cooke-Wheatstone system employed three basic states. In the former, there was no signal, a short impulse, or a long impulse. With needle telegraphy, either there was no signal or the current flowed in one of two directions. In each instance, a code system was developed from these different signals that allowed the transmission of complex information. Both the Morse as well as the needle code took the Roman alphabet as their basis; although, in needle telegraphy not all of the alphabet could be squeezed into the initial version of the code. Rather, each individual letter was conveyed by a specific combination of impulses. This method only allowed letters, numbers, and certain punctuation marks to be transmitted. All other information—such as the handwriting, handwritten corrections, or the kind of paper used, to name just a few examples—got lost in the process.

In addition, the complex encoding of letters into electric impulses led to relatively long chains of signals, even for short messages. Laying and maintaining telegraph cables was an expensive business, and for a long time, their maximum transmission loads were relatively low. Consequently, communicating by telegraph was generally a costly affair, with the price rising according to the length of the dispatch. Thus, every effort was made in telegraphic communications to achieve brevity and terseness. The information to be sent had to be to the point and free of linguistic flourishes and unnecessary digressions. Grammar and punctuation could be negotiated as desired. International business correspondence also adopted a number of special abbreviations that were set down in official code books. These codes served not to encrypt but to shorten the message in question. In the widely used ABC Telegraphic Code, for instance, the word *Aigulet* meant simply "Is not likely to affect you in any manner" (Clauson-Thue 1881: 13), and *Bluster* was "The boxes were delivered in bad order" (ibid.: 41). In this way, complex messages could be packed into a few words and transmitted in a relatively fast and cheap way. However, this also meant that every other meaning apart from the standardized content was lost. A well-known booklet containing tips and guidelines for writing telegrams puts its finger on the heart of the matter:

> Naturally, there is a right way and a wrong way of wording telegrams. The right way is economical, the wrong way, wasteful. If the telegram is packed full of unnecessary words, words which might be omitted without impairing the sense of the message, the sender has been guilty of economic waste.
>
> (Ross 1928)

This economy was important and led to the creation of a specific telegraphic style that was subject to its own socio-cultural guidelines. Brevity was generally deemed more important than correct grammar, politeness, or protocol. An interesting example can be found in a brief telegraphic exchange between the Prince of Wales and the King of Portugal marking the inauguration of the first underwater connection between Europe and India in 1870. Since the submarine cable also landed in Portugal, the prince congratulated the king in an official telegram and thanked him for the support shown by the Portuguese government. The king replied, at some remove from established procedure, "Thanks for the good wishes you expressed me in your telegram. Equally, I congratulate myself for the completion of the Telegraph. LUIZ." (British Library 1870: 21f.).[1]

Due to the perceived or real necessity of keeping telegraphic communications short and to the point, often only single, isolated pieces of information were conveyed in this manner. These were divested of any qualifying context. Only the essentials were telegraphed while other dispatches continued to be sent by letter. This is why telegraphy did not oust other means of communication (Edgerton 2006) or overwrite their spaces. Instead, the new technology fitted into an existing system of communication forms where it assumed a specific function and interacted with other media.

Instead of overcoming space or indeed annihilating it, as some contemporaries observed, (Morus 2000; Stein 1996, 2001) telegraphy produced a new, dynamic sphere of communication that existed embedded in a large number of other such spaces. The new quality of telegraphy manifested through its interactions with these changing spaces is what gave it such an important role in the context of large-scale processes such as industrialization and globalization. The following sections will lay out this new quality in the light of concrete examples in which the plurality of communication spaces and their interactions are rendered tangible through the historical figures involved. With this, light will be cast on situations in which the simultaneity of various patterns of connectivity and the numerous kinds of space present themselves with particular clarity. The concern is regarding the relationship between global connectedness and isolation, as well as, the role that telegraphy plays in this.

Fanning and Cocos: on the plurality of communicative spaces

"That sounds as if I were away down at McMurdo Sound with a South Pole expedition doesn't it!!!"—as a British telegrapher wrote in March 1914, in a letter to his friend and colleague Hollingworth in Montreal, excusing himself for the haste in which he had to pen his lines. The supply ship that took the letters was unexpectedly pushing off to sea a day earlier than planned, leaving him with just one hour in which to reply to the latest message from his friend. Almost like being at the South Pole. The postal connection, as he wrote, was simply scandalous and enough to make one despair. From this and two other letters that were donated privately to the Porthcurno Telegraph Museum it is more than clear in just what remote places in the world the telegraphers of the late nineteenth and early twentieth century had sometimes to perform their duties and just how isolated they were in many respects from the rest of the world (Porthcurno Telegraph Museum, DOC//5/107/1–3, Letters from the Fanning Islands). The letters were from an unnamed telegrapher who signed merely with the alias "Napoleon" and who was stationed in the years from 1913 to 1915 on Fanning Island. The small atoll is situated around 1450 kilometres South of Hawaii in the middle of the Pacific, and from 1889 it was part of the British Empire.[2]

Beginning in 1902, the island came to be used as a relay station in the first trans-Pacific telegraph connection. This cable run by the Pacific Cable Board linked British Columbia via Fanning, Fiji, and Norfolk Island with Australia and

New Zealand. In combination with the second cable, laid 1 year later and run by the Commercial Pacific Cable Company from San Francisco over Honolulu to Manila, the dawn of the twentieth century saw the creation of the trans-Pacific line and the closure of the last major gap in the telegraph network spanning the globe (Wenzlhuemer 2012: 118). In particular, the two Pacific cables were thus symbolic of a provisional peak in the communicative networking of the world. However, as can be seen, the staff who were stationed along this route were woven into an appreciably more complex tapestry of connectedness and isolation. Even in the middle of the Pacific Ocean, the telegraphic links between cable stations like this and the rest of the world were of course outstanding. Hundreds of telegrams were sent each day down the cables and were transcribed in the relay stations. News from all over the world was received and passed on. With that, the staff at the stations—above all the telegraphers—were always aware of world affairs. As such, they were intimately connected with the world. However, geographically many of the relay stations were extremely remote, as may be read from the letter quoted above. It was only every couple of weeks or months that a supply ship would pay a call to stations such as Fanning. This impaired personal communication with family and friends, which could only be maintained by post and not by telegraph. It also could have an effect on food supplies and necessary treatment for health problems. Looking further at this telegraph station, it will become very apparent how the various global communications spaces overlapped and how an exceptional tension arose from the relationship between these different spheres.

This becomes repeatedly clear from the aforementioned letters from Fanning Island. All in all, we have three letters from Napoleon to his friend and colleague Hollingworth. The first is dated from March 17, 1914. The second was sent roughly two months later on May 21. The third and most detailed is from January 8, 1915. Many passages in these communications convey a very palpable impression of the tension that the staff on Fanning experienced between connectedness and isolation. For instance, the author talks directly, time and again, about the circuitousness of postal communications, not only in his reference to a South Pole expedition but also in a number of other passages. Thus, Napoleon apologizes to his friend Hollingworth at the start of his second letter, saying that he has still not fully replied to his letter of November 21, 1913. He confesses to being guilty of procrastination but notes that he also has to write long letters to his wife Josephine—presumably in allusion to his own alias. "But I've had my work cut out in keeping Josephine satisfied with the promised lengthy Epistle" (Porthcurno Telegraph Museum, DOC//5/107/2, Letters from the Fanning Islands: 1). At the end of the letter, he returns to this motif (as well as to his ubiquitous allusions to French history): "If Josephine knew I have given you eleven pages she would have you guillotined!!! Am 167 pages up to her and still unfinished, so must away to do so" (Porthcurno Telegraph Museum, DOC//5/107/2, Letters from the Fanning Islands: 11).

Among other things, these ironic passages underline how scarce the postal service was from Fanning Island that linked him with the rest of the world and thus with his family. Long letters grew over the weeks and months and were then gathered together and entrusted to the supply ship. The various letters—especially those to his family—could be heavily packed and full of detail but a meaningful constructive communication was made very difficult by the sporadic manner in which the post was brought and fetched.

It was not only communication with the outside world that was inhibited under the seclusion of his post. In a number of places, Napoleon describes the precarious food situation and the scant medical attention. On occasions, when the supply ship was delayed the staff had to quickly resort to their emergency rations. In the first letter from March 1914, just such a situation is described vividly:

> Completely out of flour, therefore no bread, but we substituted with those thick square ship's biscuits which they feed the natives on, and for which I have to thank them for leaving me a Souvenir in the shape of a broken tooth! No dentist here, so what you going to do about it!!! For the first fortnight we were on famine rations. We were allowed one potatoe and one onion for dinner!!! But the last week we were completely out of these and existing on the fish we could catch, rice and these biscuits.
> (Porthcurno Telegraph Museum, DOC//5/107/2,
> Letters from the Fanning Islands: 1f.)

Likewise, in his second letter, a good two months later, Napoleon describes a very similar state of affairs:

> when our provision boat was some weeks overdue and we were reduced to famine rations!!! This time we were worse off than before.... A week before the boat arrived we were absolutely out of everything except tinned soup. No tinned meat, vegetables, fruit, milk or butter, and no fat to fry the fish in; so we had the enviable experience of getting boiled fish from Monday to Saturday.... When the last famine occurred we had a few fowls left in the run and were able to provide a poultry dinner for Sunday and an occasional egg for breakfast, but this time we had no such luxuries as that!
> (Porthcurno Telegraph Museum, DOC//5/107/2,
> Letters from the Fanning Islands: 5f.)

Even if the situation he described was not threatening to life and limb, it is nevertheless clear how fragile the station's position was in the supply network. The author's remarks on the postal communications, but above all, his frequent comments on the shortages reveal that delays in the infrequent visits from the provisions ship occurred with some frequency. In that moment, the staff was left largely to its own devices while global dispatches passed through the island in almost minute intervals.

The special characteristic of this particular interplay of connectedness and isolation is nowhere as apparent as in the third letter we have from Napoleon to

Hollingworth. In this missal dated from January 8, 1915 the writer addresses his friend after a long period of silence. Appropriately he commences with an apology:

> It is with a blush of shame (and it's a long time since I blushed) that I sit down to write you, because I have neglected you so long that you will by now, almost be wondering if I have got you out of focus as one of my friends.
>
> (Porthcurno Telegraph Museum, DOC//5/107/2,
> Letters from the Fanning Islands: 1)

Napoleon lists the usual reasons for the delay, but he also points to the change in his situation since the outbreak of war, to the enormous tension and the strenuous double shifts. The worldwide telegraph network assumed great strategic import-ance during the war, which led to the staff on Fanning Island getting involved in the events:

> But altho' we are only a mere handful of 14 Britons, we fully recognised each day, more+more, that the Empire depended on us standing up to the pressure of traffic and rendering her all assistance we could by metaphoric-ally sticking to our guns.
>
> (Porthcurno Telegraph Museum, DOC//5/107/2,
> Letters from the Fanning Islands: 3)

However, serving his home country soon went beyond this purely metaphor-ical contact with weapons, as we read in the longest part of the letter. "[Y]ou must be dying to hear a true version of the German invasion of [Fanning Island]," Napoleon writes at the beginning of his letter with reference to the arrival of the German light cruiser, SMS *Nürnberg*, at Fanning on September 7, 1914. The *Nürnberg* was part of the East Asia Squadron of the Imperial German Navy, which was officially inaugurated after the German occupation of Kiautschou in order to have a permanent naval formation to protect colonial and commercial interests in East Asia (Herold 2012; Leipold 2012; Walle 2009). After the onset of the 1914–18 war, the squadron at first mainly oper-ated in the South Pacific and attacked among other things enemy communica-tions facilities (Herold 2012: 384). It was part of these activities that brought the *Nürnberg* on September 7 to Fanning Island. The staff were in fact fore-warned. The squadron's actions were known and feared. Already a few days earlier, they had been informed by telegraph that the *Nürnberg* and the *Leipzig* had now departed from Honolulu and must be in the immediate vicinity (Porthcurno Telegraph Museum, DOC//5/107/3, Letters from the Fanning Islands). Napoleon was accordingly nervous on the morning of September 7 when a ship was detected sailing slowly toward the island. It was not initially clear whether it was friend or foe:

Personally I was of opinion that if she were a German man o' war they would commence bombarding us as soon as they reached our anchorage and … that within a few minutes we might all be blown to Kingdom Come[.]
(Porthcurno Telegraph Museum, DOC//5/107/3,
Letters from the Fanning Islands: 7)

The station superintendent viewed things differently; he was deceived by the false flags that the *Nürnberg* was flying and took it to be a French ship. The *Nürnberg* did not open fire but sent in a landing party that the superintendent, believing it to be French, welcomed with open arms. Napoleon describes in his letter how this glaring error came to light:

[T]he marines floundered out of their whaler+stood on the beach until their officer landed from the stern,+the next thing we knew was that the same officer's revolver was levelled at us with the command in a stentorian, but rather excited voice "Hands ub, you are my brizoners"! With lightning precision, of course, we all obeyed+we all fully expected to hit the beach in less than a few seconds, especially as the marines had surrounded us and stood with the rifles half way up to the shoulder in readiness to bring their weapons up upon the word of command being given.
(Porthcurno Telegraph Museum, DOC//5/107/3,
Letters from the Fanning Islands: 10)

However, the German marines had their sights set solely on communication facilities. They held the staff captive and immediately began to destroy all the technical appliances and installations they could find. They had the superintendent take them to the telegraph room. "Then the marines got busy with their axes+rifles and soon made an unholy mess of that instrument room" (Porthcurno Telegraph Museum, DOC//5/107/3, Letters from the Fanning Islands: 13). The German officers satisfied themselves that the instruments were all truly unusable and confiscated all of the code books and official papers they could find. Then came the final step:

Nothing would convince them that our engineroom was not in connection with a wireless plant on the island+they sent aboard for cases of dynamite+on its arrival proceeded to blow up the engineroom+then did the same with the two shore-ends of our cables. After blowing up the engineroom the officer was again told by someone that it was only our electric light plant+he profusely apologized+said "very sorry Gentlemen, but this is war; personally we take no delight in destruction, but we must obey orders."
(Porthcurno Telegraph Museum, DOC//5/107/3,
Letters from the Fanning Islands: 13f.)

Once all of the communication equipment was destroyed and the remaining instruments had been seized, the Germans freed their captives and departed on the *Nürnberg*. The Britons remained on the island without electricity, totally beside themselves. Napoleon gives an impressive description of how the night watches were doubled after the incident and how people forever thought they could see lights out at sea, which then turned out to be stars or some other illusion (Porthcurno Telegraph Museum, DOC//5/107/3, Letters from the Fanning Islands: 18). The mood among the staff was very tense—not least because of the total isolation resulting from the destruction of the telegraph equipment. Once the Germans had departed, they had no more contact with Fiji or British Columbia; both cables had been severed. According to Napoleon, it was also assumed that the *Nürnberg* had left underwater mines in the vicinity of the cable landing point. It took until the September 11 before someone plucked up courage to go out to the point and assess what damage had been done. The damage was considerable but at least the cable had not been mined and the station electrician could set about repairing the cable to Fiji and some of the instruments. After a few unsuccessful attempts, telegraphic communication was established on September 22 with the station in Suva on Fiji (Porthcurno Telegraph Museum, DOC//5/107/3, Letters from the Fanning Islands: 18ff.).

The relief was enormous. The station superintendent could telegraph a report on the events to Suva and the staff members could send short messages about how they were to their families around the world. On top of this, headquarters imposed a complete news embargo. Apart from a test signal every 15 minutes to confirm that the Fiji cable was operative, no messages were to be passed along the cable—neither to nor from Fanning. Since the connection with British Columbia was still down, the station on Fanning had for the time being no practical use and every message could potentially fall into the hands of the Germans; hence, the institution of a strict blackout. For the staff, this additional and seemingly unnecessary isolation was a serious blow:

> This nearly broke us up and we were all very much disgusted, but as it was the Admiralty's instructions we afterwards resigned ourselves to the bad luck with the consolation that no doubt it was for our ultimate good in baulking the enemy.
>
> (Porthcurno Telegraph Museum, DOC//5/107/3,
> Letters from the Fanning Islands: 21)

For over three weeks the men on Fanning were cut off from the rest of the world not only geographically, but also communicatively. In his letter, Napoleon claims that it was he who ultimately ended the isolation:

> It was not until 30th Sept. that Suva's observance of "Strict Silence" to us was removed by the Governor there + I fancy I was responsible for this, as, when on duty that afternoon I remarked to Suva that the monotony of this exchanging signals every 15 minutes + being forbidden any news of the outside world was worse than prison-life, and I also added that even the Nurnberg people were generous enough to give us their wireless news!!!

The fellow on duty must have felt sympathetic for us + spoken to the Supt: who, shortly afterwards interviewed the Governor + we were supplied with a morning + evening bulletin daily, after that.

(Porthcurno Telegraph Museum, DOC//5/107/3,
Letters from the Fanning Islands: 26f.)

In the three letters we have from Napoleon to his friend in Montreal between March 1914 and January 1915, the plurality and simultaneity of spaces, especially of communicative spaces, stands out vividly. Located in the middle of the Pacific, the little island of Fanning presents an almost laboratory situation in which the simultaneous inclusion of the historical actors in completely different spaces becomes especially apparent. Already, Napoleon's many remarks on the state of the provisions and the unsatisfactory postal service makes this spatial discrepancy to the simultaneous privileged access to world events tangible. Above all, it is especially present in the raid by the *Nürnberg*. On one occasion in the situation described, prior to the arrival of the German cruiser, the staff at Fanning had the latest information at their disposal concerning the whereabouts of the two light cruisers *Nürnberg* and *Leipzig*, but apart from this, they had no practical possibilities of responding to the potential threat. The simultaneity of different spaces is also revealed impressively in the moment they are suddenly balanced out by the cutting of the two submarine cables. First, through the manipulations of the Germans, then by the blackout of messages from Suva, Fanning was also cut off from the global flow of telegraphic communications. The isolation was now perfect—the geographical and the communicative spaces were virtually congruent. It is clear from the team's frustrated reactions just how strongly their insular lifeworld had distinguished itself until then by a spatial plurality that now was dashed.

There is another, no less vivid example of this plurality. In his long third letter from Fanning, Napoleon finally mentions that shortly before the connection with Suva was resumed he received a message from his colleagues on the Cocos Islands: "The next day the Staff at Cocos sent us one saying: 'Congratulations, our turn next, what!' And whether they really anticipated it or not, their turn came true enough on 10th Nov. [.]" (Porthcurno Telegraph Museum, DOC//5/107/3, Letters from the Fanning Islands: 30). Since 1901, the Cocos Islands, or more precisely Direction Island, had been home to a telegraph station belonging to the British Eastern Extension, Australasia, and China Telegraph Company—a subsidiary of John Pender's Eastern and Associated Telegraph Companies. Three undersea cables came together there: one via Mauritius and Rodrigues on the East coast of Africa, one from Batavia, and one from the Australian West coast. Since 1910, there had also been a radio station on Direction Island, meaning that the island had a certain strategic importance and came to be under attack on November 9, 1914 (and not as Napoleon writes the day after) by a German cruiser.

SMS *Emden* was removed shortly after the outbreak of the war from the East Asia Squadron and conducted its own cruiser warfare in the South Pacific and

Indian Ocean under Captain Karl von Müller. This mission took the ship to the Cocos Islands, where von Müller wanted to strike the British communications infrastructure. The captain sent in a landing force under the command of Lieutenant Hellmuth von Mücke in order to destroy all the telegraphic equipment and the submarine cable in a similar way to the destruction at Fanning. However, the *Emden* had been spotted from the island as it approached, and the staff was able to send a radio emergency signal that was received by HMAS *Sydney*. The Australian ship was close by and therefore, it was able to engage the *Emden* while the landing force was on the island; the *Sydney* inflicted such serious damage on the German cruiser in a sea battle that von Müller was forced to ground it (Wenzlhuemer 2012: 88f.).

After destroying the communications infrastructure, the landing forces under von Mücke managed to leave on a schooner, the *Ayesha*, that was anchored off of the Cocos Islands and evade capture. This curious constellation of events provided the two sides of the conflict with plenty of material to spin heroic yarns, with which both the telegraph staff as well as the landing forces were to busy themselves over the ensuing years. Like the people on Fanning before them, the staff spent a brief moment in German captivity, but according to their own accounts, they had borne this stoically without letting it spoil their mood. After the troops departed on the *Ayesha*, the electricians and telegraphers managed to repair the worst of the damage within a very short time and put at least the cable back in operation. This helped foster the myth of the manly British telegrapher like any another event. Even in the moment of great danger, the telegraphers of the *Eastern and Associated* kept a stiff upper lip and acted coolly and efficiently—at least according to the way they saw and depicted it over the years (Gagen 2013a, 2013b). The company's in-house magazine, *The Zodiac*, gave a detailed account of the incident and did all it could to boost the image of the masculine telegraphers saying,

> The men who perform these unostentatious miracles.... On desolate little islands, in remote alien cities, they lead the loneliest of lives. For conversation, they must talk across the wires to colleagues, possibly equally lonely, a thousand miles away. They know as soon as kings and mighty ones what is happening in the great world from which they are exiles; but they keep the charge with an honour as strict as their devotion.
>
> (Anonymous 1915: 62ff.)

This portrait shows the interconnections between inclusion and isolation—and with that the telegraphers' simultaneous embedment in highly differing spaces—as a central theme. However, even Lieutenant von Mücke's no less heroic reports of the incident on the island and his later journey on the *Ayesha* yield an impressive motif. After the defeat of the *Emden*, von Mücke and his landing forces embarked on a singular journey on the *Ayesha* via Sumatra to the Arabian Peninsula before finally arriving via Constantinople back in Germany, where a heroes' reception was waiting for him and his remaining men. He personally added substantially to his nimbus as hero and published two books on his adventures. In his second book from 1915, which was chiefly concerned with his quixotic return to Germany, he also described the curious meeting between the German troops and the British telegraphers on Direction Island before the *Sydney* engaged the *Emden* in an exchange of fire:

> We quickly found the telegraph building and the wireless station, took possession of both of them, and so prevented any attempt to send signals. Then I got hold of one of the Englishmen who were swarming about us, and ordered him to summon the director of the station, who soon made his appearance, – a very agreeable and portly gentleman. "I have orders to destroy the wireless and telegraph station, and I advise you to make no resistance. It will be to your own interest, moreover, to hand over the keys of the several houses at once, as that will relieve me of the necessity of forcing the doors. All firearms in your possession are to be delivered immediately. All Europeans on the island are to assemble in the square in front of the telegraph building." The director seemed to accept the situation very calmly. He assured me that he had not the least intention of resisting, and then produced a huge bunch of keys from out his pocket, pointed out the houses in which there was electric apparatus of which we had as yet not taken possession, and finished with the remark: "And now, please accept my congratulations." "Congratulations! Well, what for?" I asked with some surprise. "The Iron Cross has been conferred on you. We learned of it from the Reuter telegram that has just been sent on."
>
> (von Mücke 1917: 3ff.)

While the celebrated phlegm of the British telegraphers is themed once again in this passage from von Mücke's book, the mention of the award of the Iron Cross is especially remarkable. Hellmuth von Mücke had received the Iron Cross II class in November 1914[3] but was unable to learn as much because he was stationed on the *Emden*. The remote Cocos Islands were geographically speaking the most improbable spot imaginable to receive this information—but as chance would have it, they were perfectly integrated into the worldwide telegraph network. Von Mücke's surprise at the news, which obviously is set off in his book to dramatic effect, mirrors this discrepancy between communicative space and geographical space. As such, this frequently cited anecdote serves as yet another small example of the plurality of spaces and their meaning in the context of global history.

Space in global history revisited

The episodes discussed in the foregoing sections are particularly striking examples of the synchronicity of different kinds of space. Fanning Island and the Cocos Islands can be understood as laboratories in which the relationships and interactions between these spaces can be rendered particularly visible because in these cases the discrepancy between communicative connectedness and geographical isolation was especially pronounced as a result of the extraordinary role they played in the worldwide telegraph network. Also, outside of such remote locations, telegraphy provides a suitable context for casting light on the plurality of spaces and in particular their significance for the historical actors. This relates above all to the dematerialization of information flow via the telegraph. By separating transport and communication, new global spaces open up that work according to a completely new communication logic and are very different to already existing spaces. Examples from the history of telegraphy, such as have been described here, are thus particularly illuminating.

This principle can however be generalized. New spaces arise as part of processes of globalization or existing spaces change or assume new forms. These always relate to other spaces in which the historical actors are simultaneously involved. What is particularly interesting in this connection is the resulting changes in the relationships between the various spaces. It is these changes that give rise to new constellations, possibilities, or even limitations for the people who act in them. The impact of processes of globalization, and their meaning in people's lives and experiences become truly tangible. A complex, flexible understanding of space enables us to take a more nuanced look at such processes and with that, to be able to depart from our generalizations and from stagy metaphors of the "shrinking-of-the-world" kind. It allows us to ascertain where in the history of globalization something qualitatively new has actually arisen and what effect it had on the lifeworlds of the historical actors.

Notes

1 The omission of customary etiquette had however its limits, as is impressively shown by a few lines from a manual by Nelson Ross:

> A man high in American business life has been quoted as remarking that elimination of the word "please" from all telegrams would save the American public millions of dollars annually. Despite this apparent endorsement of such procedure, however, it is unlikely that the public will lightly relinquish the use of this really valuable word. "Please" is to the language of social and business intercourse what art and music are to everyday, humdrum existence. Fortunes might be saved by discounting the manufacture of musical instruments and by closing the art galleries, but no one thinks of suggesting such a procedure. By all means let us retain the word "please" in our telegraphic correspondence.
>
> (Ross 1928)

2 Today the atoll is named Tabuaeran and part of the island nation of Kiribati.
3 On his return to Germany in 1915, he was awarded the Iron Cross I class for his services.

References

Anonymous (1915) "The Emden's fatal visit to Cocos," *The Zodiac*, 8: 62–68.
Barton, R. N. (2010) "Brief lives: three British telegraph companies 1850–56," *International Journal for the History of Engineering & Technology*, 80 (2): 183–198.

British Library (1870) "Souvenir of the inaugural fête [held at the house of Mr John Pender], in commemoration of the opening of direct submarine telegraph with India," British Library General Reference Collection, shelfmark 8761.b.62 (June 23, 1870).

Carey, J. W. (1983) "Technology and ideology: the case of the telegraph," *Prospects: An Annual of American Cultural Studies*, 8 (1): 303–325.

Clauson-Thue, W. (1881) *The ABC universal commercial electric telegraphic code: specially adapted for the use of financiers, merchants, shipowners, brokers, agents, & c.*, fourth edition, London: Eden Fisher.

Döring, J. and Thielmann, T. (eds) (2008) *Spatial turn: Das Raumparadigma in den Kultur- und Sozialwissenschaften*, Bielefeld: transcript.

Edgerton, D. (2006) *The shock of the old: technology and global history since 1900*, London: Profile Books.

Gagen, W. (2013a) "Not another hero: the Eastern and Associated telegraph companies' creation of the heroic company man" in McVeigh, S. and Cooper, N. (eds) *Men after war*, London: Routledge, 92–110.

Gagen, W. (2013b) "The manly telegrapher: the fashioning of a gendered company culture in the Eastern and Associated telegraph companies" in Hampf, M. and Müller-Pohl, S. (eds) *Global communication electric: business, news and politics in the world of telegraphy*, Frankfurt a.M.: Campus, 170–196.

Harvey, D. (1973) *Social justice and the city*, Baltimore: Johns Hopkins University Press.

Harvey, D. (1990) *The condition of postmodernity: an enquiry into the origins of cultural change*, Oxford: Blackwell.

Herold, H. (2012) *Reichsgewalt bedeutet Seegewalt: die Kreuzergeschwader der kaiserlichen Marine als Instrument der deutschen Kolonial- und Weltpolitik 1885 bis 1901*, München: Oldenbourg Verlag.

Hochfelder, D. (2012) *The telegraph in America: 1832–1920*, Baltimore: Johns Hopkins University Press.

Holtorf, C. (2013) *Der erste Draht zur Neuen Welt: die Verlegung des transatlantischen Telegrafenkabels*, Göttingen: Wallstein.

Hugill, P. J. (1999) *Global communications since 1844: geopolitics and technology*, Baltimore: Johns Hopkins University Press.

Huurdeman, A. A. (2003) *The worldwide history of telecommunications*, Hoboken: Wiley.

John, R. R. (2010) *Network nation: inventing American telecommunications*, Cambridge, MA: Belknap Press of Harvard University Press.

Kieve, J. (1973) *The electric telegraph: a social and economic history*, Newton Abbot: David & Charles.

Lefebvre, H. (1991) *The production of space*, Hoboken: Wiley.

Leipold, A. (2012) *Die deutsche Seekriegsführung im Pazifik in den Jahren 1914 und 1915*, Wiesbaden: Harrassowitz.

Löw, M. (2001) *Raumsoziologie*, Frankfurt a.M.: Suhrkamp.

Massey, D. (1994) *Space, place and gender*, Minneapolis: University of Minnesota Press.

Massey, D. (2005) *For space*, London: Sage.

Middell, M. and Naumann, K. (2010) "Global history and the spatial turn: from the impact of area studies to the study of critical junctures of globalization," *Journal of Global History*, 5 (1): 149–170.

Morus, I. (2000) "The nervous system of Britain: space, time and the electric telegraph in the Victorian age," *The British Journal for the History of Science*, 33 (4): 455–475.

Mücke, H. von (1917) The "Ayesha": being the adventures of the landing squad of the "Emden," Boston: Ritter & Company.

Müller-Pohl, S. (2010) "The transatlantic telegraphs and the 'class of 1866': the formative years of transnational networks in telegraphic space, 1858–1884/89," Historical Social Research, 35 (1): 237–259.

Perry, C. R. (1997) "The rise and fall of government telegraphy in Britain," Business and Economic History, 26 (2): 416–425.

Porthcurno Telegraph Museum, DOC//5/107/1–3, Letters from the Fanning Islands.

Roberts, S. (2007) "Distant writing: a history of the telegraph companies in Britain between 1838 and 1868," available online at: http://distantwriting.co.uk/ (accessed October 7, 2016).

Ross, N. E. (1928) How to write telegrams properly, Girard, Kansas: Haldeman-Julius Publications.

Schlögel, K. (2003) Im Raume lesen wir die Zeit: über Zivilisationsgeschichte und Geopolitik, München: Hanser.

Soja, E. W. (1996) Thirdspace: journeys to Los Angeles and other real and imagined places, Oxford: Blackwell.

Standage, T. (1998) The Victorian internet, New York: Walker and Co.

Stein, J. (1996) "Annihilating space and time: the modernization of firefighting in late nineteenth-century Cornwall, Ontario," Urban History Review, 24 (2): 3–11.

Stein, J. (2001) "Reflections on time, time-space compression and technology in the nineteenth century" in May, J. and Thrift, N. (eds) Timespace: geographies of temporality, London: Routledge, 106–110.

Walle, H. (2009) "Deutschlands Flottenpräsenz in Ostasien 1897–1914: Das Streben um einen "Platz an der Sonne" vor dem Hintergrund wirtschaftlicher, machtpolitischer und kirchlicher Interessen" in Denzel, M. A., Dharampal-Frick, G., Gründer, H., Hiery, H., Koschorke, K., Meier, J., Pietschmann, H., Schnurmann, C., and Zeuske, M. (eds) Jahrbuch für europäische Überseegeschichte Volume 9, Wiesbaden: Harrassowitz Verlag, 127–158.

Wenzlhuemer, R. (2012) Connecting the nineteenth-century world: the telegraph and globalization, Cambridge: Cambridge University Press.

Winseck, D. R. and Pike, R. M. (eds) (2007) Communication and empire: media, markets and globalization, 1860–1930, Durham, NC: Duke University Press.

3 Using social media as a big data source for research

The example of ambient geospatial information (AGI) in tourism geography

Michael Bauder

Watch events unfold, in real time and from every angle.

(Twitter Inc. tagline 2015)

Introduction

This quote taken from the Twitter webpage makes a great promise for its users as well as the scientific community. That is, not only the communication on events will be shared with the world but also the development of events will be observable live and from every potential perspective. The promise is that we should receive a paramount and constantly updated picture of everyday and special events from the whole world, and this picture is created by the users themselves. It is a crowdsourced result of the multitude of single messages that each individual user is twittering *at* a particular place and sometimes *about* a particular place. Messages are collected under hashtags—keywords assigned to pieces of information—that aggregate to trends that, in sum, constitute events. Tweets, and similarly other contents of social media platforms, are not disconnected from space and time. Rather, next to their linguistic, audio, or visual content, Tweets possess an additional component that is of great geographical relevance; each uploaded item communicates its relation to one or many places like the place of upload, the location the picture was taken, the reference place of a message, and several time-tags. This kind of "attached" information is called ambient geospatial information (AGI), and it might be of great interest for conducting geographical research on everyday geographies.

The concept of AGI refers to geographical meta-information harvested from content that lay people contribute to online platforms including digital visual, textual, and audio data (Stefanidis et al. 2013). Especially social media such as Facebook, Instagram, Twitter, YouTube, Flickr, and others function as AGI data sources. The data shared on these platforms contain geographical meta-information such as the site and the time of both the production and upload which are attached to the actual content of pictures and texts. Thus, geographical information can be understood as a by-product of the actual platform content since they are usually not the reason why the data is originally generated and provided (Sui et al. 2013). Hence, AGI implies a non-intentional recording of geographic information in contrast to volunteered geographic information (VGI) such as data from OpenStreetMap, contributions to WikiMapia, Wikipedia, or the like, which are deliberately produced for a particular community or the public and for an ongoing use of the provided data for other purposes (Stefanidis et al. 2013). This is why AGI is sometimes called "pseudo-volunteered geographic information" (Weidemann and Swift 2013). In this sense, the AGI that is gathered, processed, and analyzed is referred to as crowdharvested, not crowd-sourced, data (Harvey 2013); this is due to the transformation of the data to scientific-useful information being done by scientific experts rather than laymen.

The geographical information on trends and events (just like the initial claim suggests), especially in the volume, velocity, and variety of social media data is a form of big data (Kitchin 2013) on topics and places with spatial and time reference and certainly do attract the interest of research. Therein, the following question is raised: which possibilities emerge from the use of this data in everyday geographies research?

In the following, this question will be discussed using the example of tourism as the field of study. Even though tourists often try to look for and experience "special" elements of any destination as well as try to distinguish themselves at any price from "typical" tourists (Edensor 2001), tourism can still be seen as an everyday practice, as many authors point out (see Haldrup and Larsen 2009; Hannam and Knox 2010; Larsen 2008a; McCabe 2005). As an everyday practice tourism could be seen as but one form of recreation activity that encompasses all activities undertaken in leisure time. It could also be seen as but one form of mobility in the perspective of the new mobilities paradigm (Mavrič and Urry 2009) within a spectrum ranging from permanent migration to daily commuting. In any understanding, tourism is cultural; "its practices and structures are very much an extension of the normative cultural framing from which it emerges" (Jamal and Robinson 2009: 3). Reflective of current societal and cultural aspects, tourism helps in the study of everyday practices. This is even more evident as more and more tourists are looking specifically for the "everyday life" of inhabitants of urban destinations (Maitland 2008).

In this regard, some answers will now be given to the question on the potential use of AGI that was posed above. First, a short overview of research on AGI in the fields of geography and tourism research will be provided. From the focus of pictures from social media photo platforms, the current stage, perspective, and epistemology of research within tourism geographies will be introduced and exemplified. Based upon this, theoretical conclusions will be derived and put into a broader framework in order to identify both potentials and pitfalls of AGI and big data-based research in tourism studies as well as geographical research.[1]

Overview on the current stage of research

The scientific use of AGI from online platforms is a relatively recent opportunity originating from the advent of the Web 2.0 in the years of 2004 to 2006 (Jackson et al. 2013). Facebook and Flickr were both founded in 2004, YouTube in 2005, and Twitter in 2006. These platforms gave way to new possibilities for internet users to share self-generated content and to create interoperability to other users and programs via so-called application programming interfaces (API) (Weidemann and Swift 2013) without the need for in-depth knowledge on programming and computers. Though the availability of extensive AGI data is a rather recent phenomenon, a comprehensive and manifold scientific research emerged quite fast (Stefanidis et al. 2013). With only approximately ten years to utilize extensive AGI data, research from a geographical perspective has focused on a plethora of topics ranging from network analyses of different social groups (e.g., Papacharissi 2009; Takhteyev et al. 2012), studies on the importance of social media for protest movements (e.g., Steenkamp and Hyde-Clark 2014; Tremayne 2014), empowerment of citizens and citizen science (e.g., Regalia et al. 2016), post-disaster processes and communication (e.g., Potts 2014), and the reliability and validity of AGI data itself (e.g., Haklay et al. 2010). Furthermore, AGI data has served as an example for Data Mining (e.g., Chareyron et al. 2013; Grothe and Schaab 2009) and the handling of big data in geographic information systems (e.g., Tsou 2014; Xu and Yang 2014).

In comparison to this variety of research interests centered on AGI, the research within the field of tourism studies appears to be far less advanced. There are some studies on evaluation platforms, such as Tripadvisor or Holiday-Check, primarily regarding images and image-formation of particular destinations (Gassiot and Coromina 2013) or economic and marketing activities (Lee et al. 2011). Furthermore, there are studies on the influence of social media on tourists and their behavior (e.g., Bizirgianni and Dionysopoulou 2013; Xiang and Gretzel 2010) as well as the variety of motivations for tourists to actively engage in social media (Munar and Jacobsen 2014). However, these studies in tourism research do not utilize the actual AGI data. Instead, they are either limited to the text and picture content or analyze the overall social media framework disregarding its content (Amersdorffer et al. 2010). In the vast overviews on the literature by Leung et al. (2013) and Zeng and Gerritsen (2014) on research on tourism and social media (and, thus, implicitly including AGI), the possibility of analyzing and integrating AGI into the research design has even been widely neglected even though there are numerous thinkable applications in researching AGI stand-alone or in combination with the picture or text content. Some research possibilities, to name a few, are analyzing and localizing tourist practices, places of photography, or even travel networks. Especially the analysis of AGI as a communicative element, as it is sent online in order to specifically communicate something, proves to be a promising research topic. For example, considering the producing/reproducing discussion, Garrod (2008) uses volunteer-employed photography, i.e., photos taken by tourists and then given directly to the researcher—as such an analog precursor of online photo data and AGI—to investigate tourists' perceptions of a destination. Using this he was able to re-open a dialogue in tourism sciences; namely, is a tourist's perception of a destination influenced by the advertisement and travel information-based communicated image and are they subsequently reproducing existing images of the destination with their photographs and experiences, or do tourists actively produce an image and thus influence advertisement and travel images by the pictures they are communicating. Garrod's research was mainly limited by the number of photos available to him. Therefore, revisiting the producing/reproducing discussions in terms of destination images and the communication of travel memories with the support of AGI could be very promising.

So far, the few works on tourism that refer to AGI data are mainly situated in the context of geographic information science and computational science (Bermingham and Lee 2014; Kádár and Gede 2013; Kisilevich et al. 2010; Orsi and Geneletti 2013). They focus on the basic viability of the use of AGI data for the analysis of the mobility of tourists or they try to model mobility patterns out of geotagged photography with reference to probability theory. To date, there have been very few works with an explicit focus on the potential of AGI data from tourism studies and especially from a tourism geography perspective. The works of Kádár (2014) and Branchet et al. (2014) appear to be an exception in this regard, although they are not transcending the stage of revealing realized mobility patterns by the use of AGI. According to the new mobilities paradigm (Cresswell and Merriman 2011; Hannam and Knox 2010; Sheller and Urry 2006), in particular the cultural meanings of practices have to be considered going beyond the analysis of mobility patterns and structures. So, though there might be examples for the application and analysis of AGI, their potential has not been estimated properly and the systematic embedding into broader (geographical) research questions has not been accomplished in most cases.

The use of AGI for mobility analysis

The basic principle of mobility analysis on the basis of AGI data is to record or harvest a sequence of geotagged photographs that are taken along a particular timeline in order to reconstruct the path that was taken by the tourist. For example, Chareyron et al. (2013) calculated the movement of tourists, or the probability of the movement to be more precise, between the sites where pictures were taken as a combination of the shortest connection between two sites and the route that is most likely taken, taking the potentially most attractive sites in the surrounding area into account. Both paths become part of the probability calculation of the distribution of paths depending upon the level of the attractiveness of the sites on the potential detour. This level is measured by the number of pictures taken and uploaded at that place. Thus, the calculated paths can be read as a measurement of probabilities for how likely it is for a tourist to use a certain path. Though such an approach may not resolve all of the conceptual limitations of conventional geographical information systems (GIS) research (e.g., Girardin et al. 2008; Kisilevich et al. 2010), it provides a much more technologically advanced research strategy. Rather than just visualizing areas of movement, this approach moves toward the prediction of precise streets. While the informatic/technological advancement lying in such an approach should not be underestimated, basic methodological questions arise in regard to what exactly is measured and calculated and what is not as well as questions on the theoretical foundation and associated understanding of mobility. In the following section the two main issues in regard to the underlying theoretical concepts are discussed.

First, the real movements of people are, as an outcome of individual perceptions and decisions, not easy to predict (Adey 2009), if predictable at all. Mobility must be analyzed in relation to the particular person based on knowledge as well as his or her individual attitudes, intentions, and motivations (Cresswell and Merriman 2011: 9). Individual movement includes the choice of a particular route that is, at the same time, an exclusion of other potential routes, and it includes, of course, a decision to move at all. These decisions depend upon individual knowledge and preferences of a person as well as other framing conditions such as the physical constitution or available infrastructure. For example, a person who becomes mobile and is visiting another city to spend his or her leisure time a few times per year (and, thus, can be characterized as a tourist in both understandings of tourism as a practice outlined above) can deliberately choose the same route every time out of habit or convenience knowing that a shorter yet unfamiliar route exists. Mobility research after the "mobility turn" (Cresswell 2011: 551) and the associated "New Mobilities Paradigm" (Sheller and Urry 2006) tries to illuminate these relations. This turn emphasizes the theorizing on the individuality of movement while, at the same time, taking developments in modern society into account—such as generally rising expectations on being mobile, and new forms of mobility and mobility patterns. The calculation of mobility paths, derived from an unspecific data basis (regarding the touristic profile of a person or a group), neglects this individual component.

Second, the upload of photographs means a *selective* representation of tourist places instead of a continuous recording of all visited tourist spots. Not every photo taken ends up on social media and not every place visited is photographed. The decision-making process and hence the process of selection is comprised of at least two steps. On the one hand, there is what Lanfant (2009: 241) characterizes as one of the pivotal questions of the "Tourist Gaze" (Urry and Larsen 2011): "Why do … [tourists] frame certain things … whereas others seem to remain 'anaesthetic'"? Addressing this problem leads to a theory of framing (e.g., Robinson and Picard 2009). That is, every picture presupposes a conscious decision (see Larsen 2008a): What do I want to shoot? What do I want to represent? What is not supposed to be shown on the picture?

> The taking of photographs involves a framing of the world; a procedure of focus, both literally and metaphorically…. The subject, be it a person or a landscape, is selected and necessarily, other people and parts of the landscape are excluded.
>
> (Robinson and Picard 2009: 13)

In this sense, taking photographs is equivalent to the photographer's accentuation of certain characteristics (Garrod 2008: 385) and photos thereby tell quite a personal story because they play an important role in capturing memories and converting them into narrations (Lo et al. 2011). For tourists, "significant places" are therefore photographed more often and in many cases are proof for having actually been at a specific sight or to having executed a certain activity. On the other hand, in the course of the upload-process all pictures are being selected with regard to a public display. The uploader's content follows from particular motivations, as analyzed by Munar and Jacobsen (2014); besides sharing special experiences of the journey, photos are uploaded to sharpen one's own profile and construct one's (online) identity. Through both selections, certain places are being over- or underrepresented compared to the actual visited spots or, to some extent, even completely ignored.

Considering these two arguments, we cannot assume the locations of photography harvested from online platforms to be an uninfluenced documentary of a tourist's mobility. Hence, the modeled movement does not reflect the real itinerary of visitors, despite the conclusion of Chareyon et al. (2013), who described online photographs as "discrete ... trace[s] of [a] real itinerary followed by a visitor." This conclusion (in small variations) is however found in many works from the computational sciences or GIScience (see also Kádár and Gede 2013; Orsi and Geneletti 2013). When taking a closer look on the line of arguments and thus the epistemologies and ontologies from these works, neopositivist positions underpinning the argumentation become visible. In his works on digital geography, Marc Boeckler characterizes these positions as "proto-naïve positivism" which have gained popularity in the context of big data research:

> In the age of Big Data there is firstly no need for hypothesis anymore, and secondly no need for causal correlations.... If the correlation of two non-connected criteria turns out to be prognostically robust, reliable statements on the course of certain events can be made.
>
> (Boeckler 2014: 9)

Transferring to the above-described setting of this article, the bulk of photo data at hand, interpreted as a tourist's unaffected, non-selected, time-dependent position, mobility pathways can be modeled through algorithms. The relationship between localized shots and the movement proves to be robust; a result is being modeled, without hypothesizing or reflecting the causal relationship from the point of view of tourism or mobility studies. The causality of a tourist's actions between arrival at a certain place, the choice of a place's elements, photographing the elements, and uploading the pictures to an online platform is left out; processes of selection which have to be situated within a particular cultural context are not being hypothesized. Therein, big data provides tourists' mobility paths that at first sight appear accurate and innovative from IT's point of view. However, the characterization of the modeled paths as a "real itinerary" seems to be purely evidence-based, not incorporating past and present theoretical aspects of mobility, tourism, and photography research, as well as their development and interconnections. Carrying out purely evidence-based research may create new knowledge but is at the same time largely ignoring earlier theoretical and empirical research, as it doesn't take these into account in the current research process. Insights just seem to emerge born from the data and are not theoretically directed and guided. Ostensibly, "correlation supersedes causation" in these approaches as the widely criticized Anderson (2008) would say with his work on and claim for the end of theory with the rise of big data.

In times of big data, positivism suddenly appears again "as just another choice in the marketplace of ideas" for research (Wyly 2014). These positions follow the tradition of data-driven social physics (O'Sullivan and Manson 2015) with its search for empirical regularities in large data sets. While data-driven research offered many prospects and influenced, provoked, and advanced geography in the past decades (Kitchin 2014a), the same is valid for big data research and in this example, AGI-based research. There is great potential in using this data if we hold on to what Kitchin (2014a) describes as data-driven geography; thus, one would avoid the showcase pitfalls outlined above:

> In contrast … data-driven science seeks to hold to the tenets of the scientific method…. It seeks to generate hypothesis "born from the data" rather than "born from the theory" …. [But] the process is guided in the sense that existing theory is used to direct the process of knowledge discovery.
>
> (Kitchin 2014b: 5f.)

Taking Kitchin's argument seriously, empirical research should acknowledge large sets of crowdsourced data, while not suspending theoretical foundations. In the given context, those foundations might be particularly consistent with a hermeneutic understanding of everyday geographies.

Opportunities for AGI-research in tourism studies

Irrespective of the critique of using AGI data to reconstruct mobility paths, it might be useful to apply these data to other questions and problems while considering the points made in the last paragraph. The potential of using AGI data is hardly assessed. This holds true especially for the combination with other methods and data, such as GPS tracking, questionnaires, image analysis, and interviews. For tourism geography, new levels of research seem to emerge with this approach, but its feasibility and validity still needs to be discussed. Through the scientific application of AGI, a late-breaking and dynamic data source is made available that is without alternative beyond big data. AGI data offers "unparalleled insight on a broad variety of cultural, societal, and human factors" (Stefanidis et al. 2013: 321), and allows us to interpret and analyze them spatially. The potential regarding AGI photo data rests primarily upon a so-called dialectics of space, resulting from the AGI data's specifics. The space the tourist encounters—or the space the tourist creates, as "performing tourism" approaches claim in accordance with de Certeau's concept of space (see Bærenholdt et al.

2004; de Certeau 1990)—is not identical to the space communicated or created by the online platform. Between these two spatialities, there are significant differences, which are characterized as follows: If a tourist encounters a tourist place, e.g., a lively square in a city with cafés, restaurants, shops, and a well-known sight, his or her perception is guided toward particular elements of it, which Urry and Larsen (2011) refer to as the Tourist Gaze. However, extending this understanding within the meaning of the performing tourism approach, each tourist not only perceives but also actively *creates* tourist spaces through his or her performance and acting with certain tangible and intangible elements of their environment (Bærenholdt et al. 2004). In this constructivist perspective a selection and negligence of elements of the space occurs as well as in perceiving them. In a next step, tourists may like to take a photo of an element or detail of the perceived space (or decides to not do so at all). Therefore, they must "frame the world" (Robinson and Picard 2009: 13) and try to make their perception and feeling of space nameable or catchable. Thereby, a photo can never capture all of the space, no matter what understanding of space is followed in the line of arguments. In addition, all "shots" of the tourist place are subject to the selection of the photographer as to whether or not they will be uploaded, as outlined above. Both the shooting and the upload of photographs to a platform are highly selective processes that can be interpreted in light of concepts of framing (Larsen 2008b; Robinson and Picard 2009), the Tourist Gaze (Urry and Larsen 2011), or of online photographs' specifics (Lo et al. 2011) as filters that turn the encountered or produced space into a told and communicated space. Therein, the space visible by others via seeing and analyzing the uploaded photos is a filtered space which cannot be recreated to its original setting and meaning. Hence, there is a difference between the space encountered and the space communicated.

The analysis of this difference in the view of these selecting processes becomes particularly valuable in answering questions about different tourist practices and places of photography, in terms of the communication of travel memories and thereby the formation of destination images, or, simply said, by all transactions between encountering and communicating everyday space and spatial experiences. One approach to these issues may be to provide a combined answer to the following two questions: "Which places were visited?" and "How do tourists represent their visited places?" This resembles a blending of actual mobility paths and the spots selected for representation with regards to space, content, or meaning and can be done through a combination of GPS-mobility data, localized photo data, and qualitative methods. The debates on the blending of visited and photographed places are currently mainly of theoretical nature (Lanfant 2009; Stylianou-Lambert 2012), if existent at all (Bauder 2016). An empirical test of the hypotheses on a quantitative level is still missing.

Further opportunities for tourism research emerge from the following aspects. The analysis of AGI data allows for research to study how spaces and places (globally, within destinations etc.) are related, or are being put in relation by tourists. Additionally, analysis of AGI data opens the avenue for research in which and how often places, elements of space, and features are picked out by tourists during and after their stay. These insights might contribute to current debates in tourism geography. For instance, AGI taken from photo data might enrich the "producing or reproducing" debate (see above; Garrod 2008) and thus discussions on spatial perception and representations of space would arise; in the case of Twitter and Facebook, news and messages could enhance our understanding of touristic dynamics after singular events (e.g., "disaster tourism," security and tourism), and many other aspects. Furthermore, the diverse elements of a destination disseminated by tourists via different social media platforms can be localized as well as put into a chronological sequence with the respective topological context through the above-mentioned combination of GPS and AGI; this provides insights into the interactions of tourists with a destination. In tourism geography, such interactions are considered a key element for the development of a tourist place or space (performing tourism; see Bærenholdt et al. 2004) and are continuously being discussed. The findings generated by the respective research might also support the practical delimitation of a city's central tourist district (CTD)—the spatially limited accumulation of tourist places and the space where tourist practices are located—as has so far been widely discussed (Duhamel and Knafou 2007), but has never been implemented convincingly. The delimitation mostly happens via the number of tickets sold at sights, museums etc., modified by the locality of other touristic infrastructure, thus giving the component of tourist practices and interactions a secondary status. The localization of photo data can provide valuable support for this delimitation.

Discussion

This contribution has shown that the use of AGI data, and data from photo-sharing platforms in particular, can be of great value for tourism research. However, compared to other branches of research, research in tourism studies still lacks a truly beneficial use of AGI data. Current AGI-research in the context of tourism is mainly conducted from the point of view of geographic information and computational science as an attempt to model mobility paths based on localized photographs. Given the dialectics of space described above, however, this approach is hardly convincing from a tourism studies perspective and lacks hypothesizing and theoretical considerations in its neopositivist approach. Big data may offer new epistemological approaches to current research questions (Kitchin 2013), but modeling without reflecting the data basis and without incorporating theoretical approaches may not be the way to generate insights into questions of everyday geographies. Given that, Kitchin (2014a) suggests data-driven science as a way to engage in big data-based research. However, what does this mean with regard to the example of AGI data? What insights may we be able to generate and how should a research project be designed according to that? The answers to these questions are complex and cannot be addressed here in full, especially in terms of research design. Nevertheless, a starting point to foster discussions will be given in the following.

The example of AGI photo data has shown that databases cannot be seen as an uninfluenced, neutral data source even in a big data context. Tempting as it may be to harvest geographical by-products of social media data and use them as an objective measure for one's own research question, the considerations above suggest that taking into account the formation conditions of the data is fundamental for any substantial and potential new insights possibly gathered. Only after analyzing the formation conditions was it possible to put the data in context and to recognize that prior to harvesting, processing, and analyzing the data, a massive selection process took place. This applies not just for online photo data but also for Twitter or Facebook data. People do not just tweet in order to send their current position to the world. Their tweets are selected to present an image. Twitter messages have a specific meaning that we must coercively analyze, and they are meant to communicate something which significantly influences what the data represent. It is tempting and comfortable to read all harvested tweets as just the current location of any person in a given area and call for a "geography of twitter" (Lansley and Longley 2016; Takhteyev et al. 2012). However, the arguments of this contribution call for a deeper look. Specifically, there is a need to fathom the formation conditions of the data as an integral part of any research design dealing with AGI harvested from social media data, and link them to social and cultural science theories for building hypotheses and examining the data in a more comprehensive way.

The insights that are potentially generated refer, on one hand, to the examined process of what the data stand for, i.e., what they represent in the first place. In the outlined example of tourism this would be the question about communicated tourist places and their relevance within the itinerary, as well as the reproduction and the sharpening of one's own (online) identity, respectively. On the other hand, the data—seen in their meaning and offspring—may allow for answering multiple questions that put the data and the things they reflect in a spatial reference. In the example of this contribution, a potential research approach was drafted that provides spatial information of visited places by adding further research methods. Utilizing both theoretical and the applied basic research, there is still the need to further investigate the suggested opportunities and show possible limitations of using social media as a big data source for researching everyday geographies.

Note

1 This contribution is an extended and revised version of Bauder, M. (2016) "Flickr, Twitter & Co: die Verwendung von Ambient Geospatial Information als Element tourismuswissenschaftlicher Forschung?" in Bauhuber, F. and Hopfinger, H. (eds) *Mit Auto, Brille, Fon und Drohne: Aspekte neuen Reisens im 21. Jahrhundert*, Mannheim: MetaGIS-Systems, 23–34.

References

Adey, P. (2009) *Mobility*, London: Sage.
Amersdorffer, D., Bauhuber, F., Egger, R., and Oellrich, J. (eds) (2010) *Social web im Tourismus: Strategien, Konzepte, Einsatzfelder*, Heidelberg: Springer.

Anderson, C. (2008) "The end of theory: the data deluge makes the scientific method obsolete," available online at: www.wired.com/2008/06/pb-theory/ (accessed May 1, 2017).

Bærenholdt, J. O., Haldrup, M., Larsen, J., and Urry, J. (2004) *Performing tourist places*, London: Ashgate.

Bauder, M. (2016) "Thinking about measuring Augé's non-places with Big Data," *Big Data & Society*, 3 (2): 1–5. https://doi.org/10.1177/2053951716665130.

Bermingham, L. and Lee, I. (2014) "Spatio-temporal sequential pattern mining for tourism sciences," *Procedia Computer Science*, 29: 379–389. https://doi.org/10.1016/j.procs.2014.05.034.

Bizirgianni, I. and Dionysopoulou, P. (2013) "The influence of tourist trends of youth tourism through social media (SM) & information and communication technologies (ICTs)," *Procedia – Social and Behavioral Sciences*, 73: 652–660.

Boeckler, M. (2014) "Neogeographie, Ortsmedien und der Ort der Geographie im digitalen Zeitalter," *Geographische Rundschau*, 66 (6): 4–10.

Branchet, B., Chareyron, G., Da-Rugna, J., Cousin, S., Michaud, M., and Pineros, S. (2014) "Observer les pratiques touristiques en croisant traces numériques et observation ethnographique: le projet de recherche Imagitour," *Revue Espaces*, 316: 105–113.

Certeau, Michel de (1990) *L'invention du quotidien Volume 1: arts de faire*, Paris: Gallimard.

Chareyron, G., Da-Rugna, J., and Branchet, B. (2013) "Mining tourist routes using Flickr traces" in Rokne, J. and Faloutsos, C. (eds) *Proceedings of the 2013 IEEE/ACM International Conference on Advances in Social Networks Analysis and Mining*, New York: ACM, 1488–1489.

Cresswell, T. (2011) "Mobilities I: catching up," *Progress in Human Geography*, 35 (4): 550–558.

Cresswell, T. and Merriman, P. (eds) (2011) *Geographies of mobilities: practices, spaces, subjects*, Farnham: Ashgate.

Duhamel, P. and Knafou, R. (2007) "Le tourisme dans la centralité parisienne" in Saint-Julien, T. and Le Goix, R. (eds) *La métropole parisienne: centralités, inégalités, proximités*, Paris: Mappemonde, 39–64.

Edensor, T. (2001) "Performing tourism, staging tourism: (re)producing tourist space and practice," *Tourist Studies*, 1 (1): 59–81.

Garrod, B. (2008) "Exploring place perception a photo-based analysis," *Annals of Tourism Research*, 35 (2): 381–401.

Gassiot, A. and Coromina, L. (2013) "Destination image of Girona: an online text-mining approach," *International Journal of Management Cases*, 15 (4): 301–314.

Girardin, F., Calabrese, F., Dal Fiore, F., Ratti, C., and Blat, J. (2008) "Digital footprinting: uncovering tourists with user-generated content," *IEEE Pervasive Computing*, 7 (4): 36–43.

Grothe, C. and Schaab, J. (2009) "Automated footprint generation from geotags with kernel density estimation and support vector machines," *Spatial Cognition & Computation*, 9 (3): 195–211.

Haklay, M., Basiouka, S., Antoniou, V., and Ather, A. (2010) "How many volunteers does it take to map an area well? The validity of Linus' Law to volunteered geographic information," *The Cartographic Journal*, 47 (4): 315–322.

Haldrup, M. and Larsen, J. (2009) *Tourism, performance and the everyday: consuming the Orient*, London: Routledge.

Hannam, K. and Knox, D. (2010) *Understanding tourism: a critical introduction*, London: Sage.

Harvey, F. (2013) "Towards truth in labeling for crowdsourced geographic information" in Sui, D., Elwood, S., and Goodchild, M. F. (eds) *Crowdsourcing geographic knowledge: volunteered geographic information (VGI) in theory and practice*, Dordrecht: Springer, 31–42.

Jackson, S., Mullen, M., Agouris, P., Crooks, A., Croitoru, A., and Stefanidis, A. (2013) "Assessing completeness and spatial error of features in volunteered geographic information," *International Journal of Geo-Information*, 2 (2): 507–530.

Jamal, T. and Robinson, M. (eds) (2009) *The Sage handbook of tourism studies*, London: Sage.

Kádár, B. (2014) "Measuring tourist activities in cities using geotagged photography," *Tourism Geographies*, 16 (1): 88–104.

Kádár, B. and Gede, M. (2013) "Where do tourists go? Visualizing and analysing the spatial distribution of geotagged photography," *Cartographica*, 48 (2): 78–88.

Kisilevich, S., Krstajic, M., and Keim, D. (2010) "Event-based analysis of people's activities and behaviour using Flickr and Panoramio geotagged photo collections" in Banissi, E. et al. (eds) *Proceedings of the 14th International conference information visualisation*, Piscataway, NJ: IEEE, 289–296. https://dx.doi.org/10.1109/IV.2010.94.

Kitchin, R. (2013) "Big data and human geography: opportunities, challenges and risks," *Dialogues in Human Geography*, 3 (3): 262–267.

Kitchin, R. (2014a) *The data revolution: big data, open data, data infrastructures & their consequences*, Los Angeles: Sage.

Kitchin, R. (2014b) "Big data, new epistemologies and paradigm shifts," *Big Data & Society*, 1 (1): 1–12. https://doi.org/10.1177/2053951714528481.

Lanfant, M.-F. (2009) "The purloined eye: revisiting the tourist gaze from a phenomenological perspective" in Robinson, M. and Picard, D. (eds) *The framed world: tourism, tourists and photography*, Farnham: Routledge, 239–256.

Lansley, G. and Longley, P. A. (2016) "The geography of Twitter topics in London," *Computers, Environment and Urban Systems*, 58: 85–96. https://doi.org/10.1016/j.compenvurbsys.2016.04.002.

Larsen, J. (2008a) "De-exoticizing tourist travel: everyday life and sociality on the move," *Leisure Studies*, 27 (1): 21–34.

Larsen, J. (2008b) "Practices and flows of digital photography: an ethnographic framework," *Mobilities*, 3 (1): 141–160.

Lee, H., Law, R., and Murphy, J. (2011) "Helpful reviewers in TripAdvisor, an online travel community," *Journal of Travel & Tourism Marketing*, 28 (7): 675–689.

Leung, D., Law, R., van Hoof, H., and Buhalis, D. (2013) "Social media in tourism and hospitality: a literature review," *Journal of Travel & Tourism Marketing*, 30 (1–2): 3–20.

Lo, I. S., McKercher, B., Lo, A., Cheung, C., and Law, R. (2011) "Tourism and online photography," *Tourism Management*, 32 (4): 725–731.

Maitland, R. (2008) "Conviviality and everyday life: the appeal of new areas of London for visitors," *International Journal of Tourism Research*, 10 (1): 15–25.

Mavrič, M. and Urry, J. (2009) "Tourism studies and the new mobilities paradigm (NMP)" in Jamal, T. and Robinson, M. (eds) *The Sage handbook of tourism studies*, London: Sage, 645–657.

McCabe, S. (2005) "'Who is a tourist?' A critical review," *Tourist Studies*, 5 (1): 85–106.

Munar, A. M. and Jacobsen, J. K. (2014) "Motivations for sharing tourism experiences through social media," *Tourism Management*, 43: 46–54. https://doi.org/10.1016/j.tourman.2014.01.012.

Orsi, F. and Geneletti, D. (2013) "Using geotagged photographs and GIS analysis to estimate visitor flows in natural areas," *Journal for Nature Conservation*, 21 (5): 359–368. https://doi.org/10.1016/j.jnc.2013.03.001.

O'Sullivan, D. and Manson, S. M. (2015) "Do physicists have geography envy? And what can geographers learn from it?," *Annals of the Association of American Geographers*, 105 (4): 704–722.

Papacharissi, Z. (2009) "The virtual geographies of social networks: a comparative analysis of Facebook, LinkedIn and ASmallWorld," *New Media & Society*, 11 (1–2): 199–220.

Potts, L. (2014) *Social media in disaster response: how experience architects can build for participation*, New York: Routledge.

Regalia, B., McKenzie, G., Gao, S., and Janowicz, K. (2016) "Crowdsensing smart ambient environments and services," *Transactions in GIS*, 20 (3): 382–398.

Robinson, M. and Picard, D. (2009) "Moments, magic and memories: photographing tourists, routist photographs and making worlds" in Robinson, M. and Picard, D. (eds) *The framed world: tourism, tourists and photography*, Farnham: Routledge, 1–37.

Sheller, M. and Urry, J. (2006) "The new mobilities paradigm," *Environment and Planning A*, 38 (2): 207–226.

Steenkamp, M. and Hyde-Clarke, N. (2014) "The use of Facebook for political commentary in South Africa," *Telematics and Informatics*, 31 (1): 91–97.

Stefanidis, A., Crooks, A., and Radzikowski, J. (2013) "Harvesting ambient geospatial information from social media feeds," *GeoJournal*, 78 (2): 319–338.

Stylianou-Lambert, T. (2012) "Tourists with cameras," *Annals of Tourism Research*, 39 (4): 1817–1838.

Sui, D., Elwood, S., and Goodchild, M. F. (2013) *Crowdsourcing geographic knowledge: volunteered geographic information (VGI) in theory and practice*, Dordrecht: Springer.

Takhteyev, Y., Gruzd, A., and Wellman, B. (2012) "Geography of Twitter networks," *Social Networks*, 34 (1): 73–81.

Tremayne, M. (2014) "Anatomy of protest in the digital era: a network analysis of Twitter and Occupy Wall Street," *Social Movement Studies*, 13 (1): 110–126.

Tsou, M.-H. (2014) "Big data: techniques and technologies in geoinformatics," *Annals of GIS*, 20 (4): 295–296.

Urry, J. and Larsen, J. (2011) *The tourist gaze 3.0*, Los Angeles: Sage.

Weidemann, C. and Swift, J. (2013) "Social media location intelligence: the next privacy battle – an ArcGIS add-in and analysis of geospatial data collected from Twitter.com," *International Journal of Geoinformatics*, 9 (2): 21–27.

Wyly, E. (2014) "Automated (post)positivism," *Urban Geography*, 35 (5): 669–690.

Xiang, Z. and Gretzel, U. (2010) "Role of social media in online travel information search," *Tourism Management*, 31 (2): 179–188.

Xu, C. and Yang, C. (2014) "Introduction to big data geospatial research," *Annals of GIS*, 20 (4): 227–232.

Zeng, B. and Gerritsen, R. (2014) "What do we know about social media in tourism? A review," *Tourism Management Perspectives*, 10: 27–36. https://doi.org/10.1016/j.tmp.2014.01.001.

4 Regionalization revisited

Mediatization of translocal social practices and the spatial reconfiguration of life in rural-urban Bangladesh

Harald Sterly

> The placelessness of communication has not only engendered fleeting functional condensation, but also the decentralization of everyday interaction.
>
> (Kopomaa 2002: 244)

> [T]he use of systems such as mobile telephony can be seen as a parallel globalization process, whereby individuals may achieve the same flexible manipulation of space and time locally as corporations have globally for many decades.
>
> (Zook et al. 2004: 168)

Introduction

The "classical" technology of electronic mediatization, the landline telephone, needed almost a century to turn *door-to-door* (where people interact face-to-face, needing to move to meet each other) into *place-to-place* communication (where people remotely interact from households, phone-shops, or offices). The mobile phone advanced the transition to *person-to-person* communication in just a few decades; in some countries this shift happened in just a single decade (Wellman 2001). The quickly emerging possibilities and practices to remotely communicate at almost any place and time, even while on the move, and on an individual, *person-to-person* basis, are changing the geographies of our everyday life; more accurately, *we are changing* the *everyday geographies* of our lives.

There is already a well-established body of literature concerning the influence mobile communication has on social relations and space, in Sociology (for example Castells 1999), Media Studies (e.g., Moores 2012), Cultural Anthropology (e.g., Horst and Miller 2006), Philosophy (e.g., Malpas 2012), Geography (see, e.g., Crang et al. 2007 or Pfaff 2010 for an overview), or in Development Studies (e.g., Donner 2008 or Duncombe 2011 for extensive reviews). Therein, what should be the added value of another text on this subject? There are two reasons to add something here.

First, there seems to be an imbalance of research focus; most studies on broader issues of social and cultural change (e.g., geographies of everyday life) are located in the context of "industrialized" countries. In the context of "developing" countries, the larger part of research is concerned with more practical topics, such as economic value chains, m-health (for *mobile*), m-banking, m-information, or m-governance; from the early 2000s onwards, the distinct, inter- and transdisciplinary research field of Information and Communication Technology for Development (ICT4D) has been emerging. There are also notable exceptions (e.g., Kleine 2010, on the freedoms of choice and the scope of agency that ICT might entail; Horst and Miller 2006, on the entanglement of mediatized communication, culture, and social practices in Jamaica; Paragas 2009, on the spatio-temporality of overseas Filipino workers' family relations). However there remain obvious disparities of research foci in "developed" and "developing" contexts. This is understandable, given many structural problems in developing contexts (and the expectations in the mobile phone for solving them), but with the global rise of mobile communication (and a convergence of mobile penetration rates), also developing countries "deserve" more attention regarding broader aspects of social change.

The second reason is that the mobile phone marks a significant transition in the way that migrants—and this applies also for hundreds of millions of labor migrants—live their everyday lives *over a distance*, spanning places, countries, and continents. There are relatively few studies in the nexus of these two fields, explicitly linking translocal livelihoods and mobile communication, for example Miller (2009), Brickell (2011), or Tan and Yeoh (2011).

Thus, this chapter follows two aims. First, contributing to knowledge on broader social and spatial implications of mediatization—the (re)making of everyday geographies—in a developing context (Bangladesh). The second aim is to abstract from this and to contribute to an empirical and conceptual body on the linkages of mediatization[1] and translocal living. In the following section the chapter will give a brief overview of the setting of this study in Bangladesh and the respective underlying theoretical framework and methodology. It will then illustrate the relevant processes of change in communication and spatial relations, with empirical examples, and conclude with a reflection on what this implies for spatial and social processes.

Setting the scene: mobile communication and translocality in Bangladesh

In the roughly four decades since its inception in 1978, the global number of mobile-cellular subscriptions has almost reached the number of global population; in 2015, there were 7.2 billion SIM cards with 7.3 billion people on earth (ITU 2016a; UNDESA 2016). As a result of this rapid diffusion, the mobile phone was dubbed as the fastest spreading technology in human history (Worldbank 2012). Bangladesh is no exception to this, with a number of 134 million subscriptions in 2015, equalling a "penetration rate" of 83 percent of the 160 million people living there (ITU 2016b). Although differential access (the "digital gap") continues to be an issue, especially when it comes to mobile data and accessing the internet, the number of Bangladeshis without access to basic forms of mobile communication has been, and is continuously, shrinking. A representative survey conducted in 2015 in Bangladesh found that 64 percent of the respondents owned a mobile phone but 96 percent actually used one (FII 2016). While Bangladesh is often covered in general reports on ICT (information and communication technologies) and development, specific research on mobile communication in Bangladesh is relatively scarce; some examples are Bayes (2001) on the village phone initiative, Bhuiyan and Alam (2004) on telecommunication policies, Rashid (2011) on phone-sharing in rural areas, Hossain and Beresford (2012) on gender issues or Islam and Grönlund (2011), and Dey et al. (2013) on phone use by farmers.

Regarding translocal livelihoods, Bangladesh is characterized by high population mobility, both international as well as domestic. While international migration is relatively well documented, there is only sparse information on domestic migration figures. It is, however, observable in the forms of both the rural-to-urban movement of permanent, temporary, or seasonal labor migrants from the poorer North, Northeast, and South to the large urban areas of Dhaka and Chittagong, as well as the respective urban growth rates. Since independence in 1971, the economic, political, and symbolic primacy of the new capital, Dhaka City, has attracted millions of people, resulting in urban population growth rates of up to 10 percent (Siddiqui et al. 2004; United Nations 2012). Compared for example to China, rural-to-urban migrants in Bangladesh tend to keep strong relations with their places of origin, reflected in frequent visits and more persistent up keeping of translocal connections (Bork-Hüffer et al. 2016). The many million migrants, who are occupied in the garment and construction industry, as rickshaw pullers and street-food vendors or as domestic workers in Dhaka's richer households (Siddiqui et al. 2010), maintain relations with their households or families in thousands of villages and towns all over Bangladesh, thus connecting these places through networks and flows of people, information, finances, and goods.

In this context, ICT and mediatization of communication, associated with the abilities to communicate over distances and in real time, can be expected to considerably influence the ways in how migrants and their distant family members live and practice their relations over distance. The causal linkages between changes in translocal livelihoods and the mediatization of communication are, however, far from simple.

Conceptual framework, methods, and data

Translocality here refers to structures and practices that link people and places over distance, encompassing both mobile (migrants) as well as non-mobile (migrants' families at their home place) populations (Brickell and Datta 2011). It is along such relations, motivated by imaginaries and aspirations, attachments and notions of belonging, that translocal *flows* of people, finances, knowledge, and goods are established and maintained. I build upon a conceptualization of translocality proposed by Greiner and Sakdapolrak (2013), as constituted by the three aspects of place, networks, and locales. They underline the co-constitution of places and networks (as structures) on the one hand and agency on the other, based upon Giddens' Theory of Structuration (Giddens 1984). For my conception of translocality, this implies the consideration of places, relations, and institutions as having structural properties that enable and constrain agency but are also outcomes of agency (see the left and central part of Figure 4.1). Two important elements of the framework will be sketched briefly: *position-practice-relations* and *place*.

Position-practice-relations

Similar to the concept of networks, position-practice-relations (PPR) also capture the relations between people. However, whereas networks refer to the relations between *concrete individuals*, PPR refer to the generalized relations between *categories of actors* (positions) and the related roles and practices (Bhaskar 1979). Following Cohen (1989), four core aspects constitute PPR: (1) *Positional* identities, associated with entitlements and obligations, as well as access restrictions, such as qualifications, age, or kinship. As pointed out above, positions are understood here as *categorical social identities*, e.g., as professor, daughter, or landlord, and not the actual persons instantiating these positions. (2) Sets of *practices* associated with or expected from a given position. Practices refer to socially structured and routinized ways of doing things—of thinking and understanding, communicating, using the body, but also of handling objects (cf. Reckwitz 2002). A professor is, for example, expected to research and publish, to teach and supervise, and to engage in departmental affairs. (3) The *relations* with which a given position-practice is linked and embedded (and thereby also defined) in a web of other position-practices; a professor would, for example, be

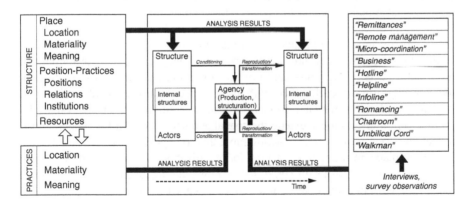

Figure 4.1 Conceptual framework and analytical procedure. Left part of the figure: theoretical framework of translocality and place; middle part: relations and dynamics of structure, actor, and agency in the process of structuration; right part: types of mediated communication practices derived from observations and interviews; thick black arrows: analytical procedure.

linked to discipline and departmental colleagues, students, or the university's administration, and through these linkages, the position of a professor is largely defined. (4) These relations and practices, for example the specific ways that the professor interacts with other positions (e.g., colleagues, students, etc.), are *institutionalized*, thus enduring through time and often spread over space (Cohen 1989: 210).

The concept of translocality implies a multidimensional, dynamic, and essentially relational notion of place (Brickell and Datta 2011; Greiner and Sakdapolrak 2013). I distinguish three major dimensions of place, following Agnew (1987): *Location*, as "the geographical area encompassing the settings for social interaction" (Agnew 1987: 28); the *material* sites in which agents are corporeally present and co-present, interact, and where material resources are situated; and the *meaning*, including actors' ways of making sense and identification with a place, as well as the social norms, values, and rules that are particular to a specific place (Sterly 2015). Places are constituted through both local and *relational* structures and processes—internal arrangements (e.g., location, environment, architecture, traditions) are always complemented by relations to other places (e.g., transportation infrastructure, trade networks, flows of people and goods, expectations, etc.). It would be of little help for understanding a specific place if only its local *or* relational/translocal structures and practices would be considered (Massey 2005). These structural aspects of PPR and of place do condition—enable and constrain—agency and practices of social actors in and between places but are in turn themselves influenced by them; thus, they are also outcomes of agency (see middle part of Figure 4.1).

PPR, as well as interactional situations, are to a large extent linked to places; they are *regionalized*. It is usually only appropriate to enact certain practices (and thus "enact" the position), when in the respective places, at the respective times, or in the presence of the respective others; a professor and a student have specific interactional practices bound to their positions during an exam that are different from the interactional practices required or expected in a possibly even bodily interaction during a departmental soccer match and on the sportsground. The hierarchical differences, repertoires, and scopes for agency, as well as the formal and informal aspects of individual and positional relations, depend not only on their positionality toward each other, but also on the spatio-temporal setting (the "locale") and the situational context. Mediated communication and especially mobile communication can extend the spatial and social context for enacting positional practices; for example, enacting more than one position at a given time and at a given place, resulting in a potential for pluralization of communicative or positional contexts. The professor could, for example, talk to colleagues or students on the phone when queuing in front of the supermarket cashier.

Methods and data

The data for this study was obtained through fieldwork in Dhaka and two villages North of Bangladesh between 2011 and 2013. Methods for data collection included observations, expert interviews, semi-structured interviews with villagers and rural-to-urban migrants, focus group discussions, and a survey among 145 village and 46 migrant households. The interviews were recorded, transcribed, and translated, and were then analyzed with qualitative content analysis using the software MAXQda.[2] The quantitative survey data was analyzed in Microsoft Excel. From the observations and interview data, 11 types of frequently occurring mediatized communication practices were inductively derived (see Figure 4.1, right part) and the interactions with structural translocal aspects were analyzed (see Figure 4.1, thick black arrows).

Case examples

In the following section, the linkages between socio-spatial changes and mediatization of communication will be explicated, using the illustrative examples of three of the mediatized communication practices derived from the interviews and observations: "remittances," the replacement of remittance sending via personal couriers with sending through mobile financial services (MFS); "romancing," the emergence of a new form of mediatized intimate (extra-marital) relations; and "chatroom," the frequent calling of distant family members and friends to exchange news and experience a sense of proximity. Table 4.1 gives an overview of the structural outcomes of these types of mediatized communication practices.

Remittances

Before the widespread use of mobile phones, money was sent with couriers, for example family members, friends, or other trusted persons, or it was posted via formal or informal money-sending services:

> If I had no mobile, I could not have talked for a single day. Someone might go to the village, then I could send money or message through him. Then he will go and tell them. Before that I can't inform them how much money I have sent. But now if I send 500 taka, I can inform them within one minute that I have sent 500 taka.
>
> (Kazi, Rickshaw puller, Dhaka, interview November 6, 2011)

Now, via mobile phone, information is exchanged about the remittance process (communicating financial demand, and organizing and monitoring the procedure of the transfer) and the money is increasingly sent directly through MFS. At the time of the interviews (2011–2013), the majority of the interviewees used mobile phones to exchange information; "Suppose, think, I have gone to Dhaka. At home they have no way for eating. [They] need money. On phone they would say that 'I need money, send me money'" (Shahin, rickshaw puller, Interview Rangpur village, November 15, 2011). Only about 5 percent of the interviewed used MFS, a figure that has presumably increased significantly, given the overall growth of the

Table 4.1 Structural outcomes of mediatized (mobile) communication practices

Structural aspects of translocality	Position-practice relations		Mediatized (mobile) communication practices		
			"Remittances"	"Romancing"	"Chatroom"
	Positions, Positionalities		Organized system: new "nodes" (MFS agents), more security for clients, saving of time and money, increasing importance of MFS agents	Increased scope of agency (to overcome constraints)	"Peace of mind," (more control over women)
	Relations		New and formalized relations of clients to MFS agents	Easier to engage with intimate partners: more and denser (illicit) relations	Intensification of relations (core networks), conflict resolution
	Practices and Institutions		Ease of remitting and control: more frequent sending, (less physical mobility of couriers)	More frequent interactions, (rising acceptance of non-marital relations)	"Quotidianization,"[3] increased expectations for more frequent calls
	Place	Location	New places, re- or decentralization	New places for intimacy (fields, workplaces, roads, etc.)	Distance feels less important
		Materiality	Grocery stores and phone shops upgrading, advertisement	Privacy of places important for secret interaction	—
		Meaning	New translocal meaning of formerly "predominantly local" places (tea stalls, grocery shops, etc.), increasing importance of these places	Places become locales for distant everyday practices of intimacy and romance; possibility for women to escape isolation	Any place becomes potential focal point for chatting; "doubling of place"

MFS market in Bangladesh since its inception in 2011 (FII 2016). The transfer of money via MFS is easy and quick; sender and recipient have to register with a service provider (banks and phone companies), the sender deposits cash with an MFS agent, and the amount is credited on the recipient's balance. The recipient can then withdraw the amount (or less) from any MFS agent of the same provider. The transfer is also done without registration; in these cases, the MFS agents act as (informal) intermediaries, charging more than the official rate.

These shifts imply major structural changes in the topology and spatial configuration of remittance flows; a whole new functional system is emerging with MFS agents forming the new material nodes in the translocal financial relations. They are often micro-entrepreneurs (for example, mobile recharge, grocery shops, or tea stalls), and this extension of their service increases their importance. The relations between users and service providers are becoming more formalized—fees are transparent and equal for everyone, and the transfer is legally secured—which also changes the positionality of users. Interviewees highlighted the convenience of the MFS for remittances, with reduced costs, time spent for organizing the process, and insecurity; "Which one [transaction mode] do you use most the time now, which media?"—"bKash [one brand of mobile financial services] is the most convenient" (Forid, factory worker, Dhaka, interview March 9, 2013). This also contributes to more frequent remittance sending:

INTERVIEWER: "Earlier, how many times was it [money transactions] earlier?"
FORID: "Once [per month]"
INTERVIEWER: "And now approximately?"
FORID: "Twice or thrice (laughs)" (Forid, factory worker, Dhaka, interview March 9, 2013).

The spatial changes in remittance flows can be interpreted as being centralized, compared to the personalized and individual money delivery; in 2013, the majority of the approximately 60,000 MFS agents (FII 2016) were located in smaller towns and market centers, the closest agents in the study area being approximately 3–5 km away from the two villages. The further spatial spread of MFS services since then has led to a decentralization again; the number of MFS agents has increased to approximately 180,000 in the 56,000 villages and settlements of Bangladesh (Parvez et al. 2015; BBS 2016), and two-thirds of the population now has access to MFS agents within 1 km of their home (FII 2016). Advertisement for MFS services—and sometimes also improvement of the physical structures—of village grocery stores or tea stalls are the visible symbols and material signs of the new translocal features of these formerly rather local places, reflecting also their changed meaning and importance in the village topology.

Chatroom

The practice type of "chatroom" represents communicative situations where (distant) actors call each other to experience co-presence and shared time as well as to exchange information on everyday issues, such as their wellbeing or the state of affairs at the distant place (mostly the village). The description "chat-room" is inspired by Anwar, a rickshaw puller in Dhaka, who describes the following practice:

> Going out [in Dhaka] I get 20 Taka balance in mobile and talk with friends [in the village]. They are sitting down there, switching on the loud speaker, four to five people listen to me and have fun.
> (Anwar, rickshaw puller, interview Dhaka, March 22, 2011)

The most apparent feature in terms of practices is the quotidianization of such interaction—communicating with family and friends—changes from being something extraordinary to an everyday practice.

Frequent, often daily, chats with family members and relatives are important to most of the interviewees, as it means an improvement in psychological well-being and contributes to strong and reliable social relations. A frequently occurring motif in the interviews was that the daily communication—and the possibility to call distant family members at any time and from any place—reduces "tension" and contributes to "peace of mind." Abdul Aziz, who lives in Dhaka and has his elderly parents in the village, explains the difference that mobile communication compared to writing letters makes to him:

> Letter, it is like that.... It can't go the day we send. So, in mobile we can listen the news in every minute. A letter needs seven days ... to go any-where. ... No, no, sending letter is not [a] peaceful thing. But through mobile I can hear the voice of mother.... From here today I write a letter, they will get it after three days. Then the parents will return the letter, is it a matter of peace? It is unpeaceful.
> (Abdul Aziz, rickshaw mechanic, Dhaka, interview November 1, 2011)

Daily interactions also means more control over farming and other village household decisions by the absent male household head, including control over women's whereabouts:

> [Women] should have a phone because I can reach them all the time to talk.... I can learn about everything, when there is any problem at home, I can know.... I hear everything from there and if there is a conflict with anyone [I can] hear that over phone.
> (Khokon, Rangpur village, interview November 14, 2011)

The majority of the interviewees experienced the daily conversations as improving the relations with their distant relatives:

> For the mobile phone, now the relationship is better. (Interviewer: Why?) [The] communication process is good. Now the communication with Dhaka

is the matter of just a minute.... Communication with relatives, friends, for business, all are much better. Very beneficial.
(Helal, businessman and teacher, Rangpur village, Interview March 8, 2013)

As the "chatroom" via mobile phone is entered especially with family, relatives, and friends, this specific practice contributes to an intensification of "core networks." This is in line with other findings on the effects of mobile communication on network patterns in India, Kenya, and the USA (Hampton et al. 2011; Palackal et al. 2011; Shrum et al. 2011).

Increased communication frequency and intensity can, on the other hand, also lead to an increased demand for communication. Forid illustrates how distant relatives expect more frequent communication since they have mobile phones:

[A]t that time [without mobile phones] people would understand that the situation was like that [they used to talk every 2–3 months], that it would take time to communicate ... but after getting the mobile, talk every 2 days or every day. If we don't talk now they don't take it positively, they think "they are not talking" [with us].... Now if we don't talk two days, they will call (Interviewer: okay okay) to ask "how are you?"
(Forid, factory worker in Dhaka, interview March 9, 2013)

Daily communication influences the perception of distance: "If I communicate I feel that I am able to get the news from home every day, don't feel the distance" (Forid, Factory worker, Dhaka, Interview March 9, 2013). Foyez, a village farmer with children working in Dhaka, explains: "[F]or using mobile it seems like they are living nearby, we can contact anytime ... we become closer by using mobile" (Foyez, farmer, Rangpur village, interview March 8, 2013). Given the portability and personalization of communication afforded by mobile phones, any place can become a locale for a "chatroom." The local anchoring of translocal communication is influenced by preferences, but also by phone ownership; while non-owners have to use phones of relatives or friends or the services of call shops, phone owners can call from more private places such as their homes. At home, the phone can also be handed over to other family or household members present during a conversation. However, calls are also made and received from public places such as roads and foot paths, open spaces, tea stalls etc.—implying frequent overlaying of different communicative contexts ("doubling of place," cf. Moores 2012).

Mobile romancing

In Bangladesh, the term "mobile romancing" commonly refers to the calling of others' (mobile) numbers and attempting to enter a romantic relationship or to sustain such a relationship through the phone (cf. Sterly and Gerads 2016). The numbers of contacts are sometimes dialled randomly; sometimes they are given by friends or are acquired from prepaid balance vendors. In many cases, this is perceived as harassment, especially by women, but in some cases, this results in longer-lasting relationships. One interviewee explains:

> Suppose a call or message comes in someone's mobile. In fact there is no relation or identification of that girl. Suppose a boy has got a number of a girl and she is from some [other] place, suppose from Dhaka to Rangpur. "From where have you got my number?" "From that person." In that way people talk in an unknown number.
>
> (Interview with Bimol, rickshaw puller in Dhaka, interview November 6, 2011)

The interviewees—mostly men, who answered on the topic directly—experienced mobile romancing generally as positive; for instance, they describe it as a diversion from their daily life. One interviewee said: "… As I'm a man, it's a matter of manly emotion. It feels good to talk with other [women]. But it's not possible to talk more freely in face to face, that is easier in mobile" (Maruf, farmer, interview in Rangpur, November 16, 2011). The survey—although not representative—indicates that this practice is not a marginal one; one out of ten surveyed villagers and seven out of ten interviewed migrants in Dhaka reported having experienced mobile romancing before. In general, the practice of mobile romancing and the (mostly extra-marital) relations established are regarded as illegitimate, and the mobile phone helps to keep the practices and relations secret.

Actors actively employ the affordances of mobile communication—translocality, personalization, portability—to overcome some of the rigid constraints regarding non-marital romantic relations and intimacy and to conceal their relationships. This can be interpreted as an increase in actors' scope of agency and thus, as a certain shift of positionalities, especially those of women and adolescent girls who face distinct limitations in terms of social and spatial mobility (Sterly and Gerads 2016). Many interviewees opposed adolescent girls having access to mobile phones; one said: "[S]uppose if girls have mobile, [and the] girl is not married yet … that girl build a relationship with a boy, is not it bad? … Don't we have to control her, don't we have to bind her in an environment" (Anwara, housewife, Rangpur village, interview November 12, 2011). Through engagement in romantic relations over the phone, new types of relations are established that could not, or have only been possible to a certain extent, without the mobile phone. The transgression of social rules regarding how romantic or intimate relationships should be established and practiced is likely to also contribute to changes in these rules over time. The fact that there is the known and established term of "mobile romancing" in Bangladesh indicates at least that it has entered public discourse and consciousness as a social practice.

Mobile romancing is practiced between and at places that might not have been associated with intimacy and romance, for example workplaces, dormitories, roads, or public places. Among students of the public universities in Dhaka, it is common to call their intimate partners at night from their dormitories and have conversations that can last for several hours. One interviewee said: "I am calling most of the time from [the] department, but on midnight I call her, then calling 1 to 2 hours, 1 am to 3 am ..." (Student Interview 2, male, Dhaka, March 20, 2013). Maruf, a village farmer, tells the story of a relative who had an extra-marital relation over the phone for many months with a woman living in the same housing compound. He said that the relative

> used to talk with that girl for hours.... Sometimes one of them was in duty [calling from there].... Although they used to live in the same flat, they used to talk through mobile phone.... In fact, they couldn't get a chance to meet directly in the house.
>
> (Interview with Maruf, November 16, 2011, village in Rangpur)

For such evasive practices, the privacy of places is of course highly relevant, either through walls, solitude, or anonymity. Through such practices, places acquire also new meanings. Students' dormitories, for example, have become hubs of romantic night talks and secret love affairs—in addition to their original purpose.

Conclusion

These examples illustrate three underlying processes of socio-spatial change that are associated with mediatization: a translocalization, a re-regionalization, and a shift in power relations. They relate to, but also go beyond the themes that Julia Pfaff (2010: 1435ff.) identifies in geographical research on mobile communication: "presence-absence," "public-private," and "freedom-control."

Spatially distant actors are *translocalizing* the topographies of their *interaction*—before they communicated over the mobile phone, their communication was largely taking place locally, being limited to the occasional face-to-face meetings (with the exception of letter writing, which was not very common among poorer rural-urban labor migrants). With their frequent calling (and the ability to do so), the occasional and extraordinary interaction becomes every day, quotidian practice. The experience of connectedness, the impression of co-presence, and of the simultaneous "taking part" in the daily life at *the other place* adds a fundamentally translocal dimension to social life of many rural-to-urban migrants. The costs that poorer migrants bear for that indicates that they value this *quotidianization* of proximity. What Pfaff formulates as "the experience of being there and not being there," and as "mixtures of absence and presence [that] arise through the use of mobile phones" (Pfaff 2010: 1436)—and what one interviewee puts as "don't feel the distance"—matches well with an understanding of translocality that explicitly aims at capturing the "simultaneous situatedness across different locales" (Brickell and Datta 2011: 4).

However, through mediatization, these practices (remitting, romancing, and chatting) do not only take place between distant places; the spatial structure of remittance flows is being reconfigured, and intimacy and proximity are literally taking new places, being expanded to places beyond rural and urban home-steads, to paths and roads, factories and offices, buses, and rickshaws. I inter-pret this as a *re-regionalization* of practices; it is a restructuring of the spatial and temporal ordering of everyday life, in the sense of Giddens (1984) and Werlen (2009). This is also linked to a pluralization of place, in the sense of the overlay of co-present and mediated communicative situations (Moores 2004). An example is an actor engaged in a private or intimate conversation in a public place and the others that are co-present actors at that place. The effects of this pluralization or hybridization of place on the relations between *public* and *private* realms have been a prominent subject of geographical and sociological studies (Pfaff 2010).

Although the empirical data does not allow for a generalization of statements about changes in power relations, the above examples indicate an increase in the actors' capabilities to overcome constraints and a widening of their scope for agency. However, the cases also show that increasing *freedom* comes with increased potential for *control*; when sending remittances through MFS, the extended possibilities contrast with increasing expectations from the villagers toward the migrants to send money more frequently. The mediatized long-distance chatting enables the creation of proximity and gives a sense of security and peace of mind. However, it also permits tighter control of affairs by husbands and household heads. For those engaging in mobile romancing—especially adolescents and women—the ability to subvert and transgress strong normative constraints and restrictions regarding mobility and intimacy is most notable. At the same time, there are also aspects of tighter control of the where-abouts of daughters and wives; here, the mobile phone functions both as a "port-able purdah" (enabling more mobility and freedom; see Sterly and Gerads 2016) and an "electronic leash" (Caron and Caronia 2007: 210), enabling control and supervision.

There is one important point to make here: mobile communication is hap-pening *not independent* of place, as that would imply a *de*-regionalization of communicative practices that I do not see happening here. The portability of mobile phones and the reduced dependency on fixed communication infra-structure at places rather increases the relative importance of other aspects of places, aspects that qualify them for engaging in communicative practices; Examples include: material (walls affording privacy, sitting facilities, noise level), social (others around; for example, teachers or parents), and normative or identity-related aspects. Mediated communication does not necessarily lead to the substitution of (or stimulation of additional) face-to-face communica-tion, but rather, it complements the latter and "a complex co-evolution, articu-lation and synergy between place-based and telemediated exchange" emerges (Graham 1998: 172).

Thus, to take up Zook's statement from the beginning of the chapter, what we can observe in Bangladesh—as in many other parts of the world—is a form of "mediated globalization from below." "Ordinary people," who would not com-monly be regarded as core agents of globalization, are actively employing affordances of mediated communication in order to establish and intensify relations to other people and places, across scales and boundaries, and are thereby reworking—translocalizing and re-regionalizing—the topographies of everyday life.

Notes

1 With regard to mediatization, I focus on one particular aspect only—the rise of mobile communication, with an emphasis on voice communication. Mediatization, in general, encompasses a broader field of communication technology, including social networks, virtual reality, etc., as well as the *convergence* of technologies, e.g., mobile internet applications. I argue that, although limited, this focus makes sense in the case of the study region and group (rural-to-urban labor migrants in Bangladesh), where between approximately 2000 and 2015 access to cheap mobile phones and affordable calling rates meant the onset of mediated communication for large parts of the population.
2 MAXQda is a software package for qualitative and multi-method analysis of texts and multimedia content.
3 *Quotidianization* refers to the change from the extraordinary and exceptional to the everyday quality of practices; it is a translation from Max Weber's (1922) *Veralltäglichung* via the French *quotidiennisation* (Bégout 2005; Schouten 2012).

References

Agnew, J. A. (1987) *Place and politics: the geographical mediation of state and society*, Boston: Allen & Unwin.
Bayes, A. (2001) "Infrastructure and rural development: insights from a Grameen Bank village phone initiative in Bangladesh," *Agricultural Economics*, 25 (2–3): 261–272.
BBS Bangladesh Bureau of Statistics (2016) *Statistical year book Bangladesh 2015*, available online at: http://203.112.218.65/WebTestApplication/userfiles/Image/SubjectMatterDataIndex/YearBook15.pdf (accessed December 12, 2016).
Bégout B. (2005) *La découverte du quotidien. Éléments pour une phénoménologie du monde de la vie*, Paris: Allia.
Bhaskar, R. (1979) *The possibility of naturalism: a philosophical critique of the contemporary human sciences*, Atlantic Highlands: Humanities Press.
Bhuiyan, A. J. and Alam, S. (2004) "Universal access in developing countries: a particular focus on Bangladesh," *The Information Society*, 20 (4): 269–278.
Bork-Hüffer, T., Etzold, B., Gransow, B., Tomba, L., Sterly, H., Suda, K., Kraas, F., and Flock, R. (2016) "Agency and the making of transient urban spaces: examples of migrants in the city in the Pearl River Delta, China and Dhaka, Bangladesh," *Population, Space and Place*, 22 (2): 128–145.
Brickell, K. (2011) "Geographies of "home" and belonging in translocal Siem Reap, Cambodia" in Brickell, K. and Datta, A. (eds) *Translocal geographies: spaces, places, connections*, Burlington: Ashgate, 23–38.
Brickell, K. and Datta, A. (2011) "Introduction: translocal geographies" in Brickell, K. and Datta, A. (eds) *Translocal geographies: spaces, places, connections*, Burlington: Ashgate, 3–22.
Caron, A. H. and Caronia, L. (2007) *Moving cultures: mobile communication in everyday life*, Montreal: McGill-Queen's University Press.

66 *H. Sterly*

Castells, M. (1999) *The rise of the network society, The information age: economy, society, and culture Vol. 1*, third edition, Malden: Blackwell.
Cohen, I. J. (1989) *Structuration theory: Anthony Giddens and the constitution of social life*, Basingstoke: Macmillan.
Crang, M., Crosbie, T., and Graham, S. (2007) "Technology, time-space, and the remediation of neighbourhood life," *Environment and Planning A*, 39 (10): 2405–2422.
Dey, B. L., Binsardi, B., Prendergast, R., and Saren, M. (2013) "A qualitative enquiry into the appropriation of mobile telephony at the bottom of the pyramid," *International Marketing Review*, 30 (4): 297–322.
Donner, J. (2008) "Research approaches to mobile use in the developing world: a review of the literature," *The Information Society*, 24 (3): 140–159.
Duncombe, R. (2011) "Researching impact of mobile phones for development: concepts, methods and lessons for practice," *Information Technology for Development*, 17 (4): 268–288.
FII Financial Inclusion Insights (2016) "Report on the Bangladesh wave report FII tracker survey, conducted August–September 2015," available online at: http://finclusion.org/uploads/file/reports/2015%20InterMedia%20FII%20BANGLADESH%20Wave%20Report.pdf (accessed February 12, 2016).
Giddens, A. (1984) *The constitution of society: outline of the theory of structuration*, Cambridge: Polity Press.
Graham, S. (1998) "The end of geography or the explosion of place? Conceptualizing space, place and information technology," *Progress in Human Geography*, 22 (2): 165–185.
Greiner, C. and Sakdapolrak, P. (2013) "Translocality: concepts, applications and emerging research perspectives," *Geography Compass*, 7 (5): 373–384.
Hampton, K. N., Sessions, L. F., and Her, E. J. (2011) "Core networks, social isolation, and new media," *Information, Communication & Society*, 14 (1): 130–155.
Horst, H. and Miller, D. (2006) *The cell phone: an anthropology of communication*, Oxford: Berg.
Hossain, S. and Beresford, M. (2012) "Paving the pathway for women's empowerment? A review of information and communication technology development in Bangladesh," *Contemporary South Asia*, 20 (4): 455–469.
International Telecommunication Union ITU (2016a) "Global mobile-cellular subscriptions, table 'Mobile Cellular World'," available online at: www.itu.int/en/ITU-D/Statistics/Documents/statistics/2016/Stat_page_all_charts_2016.xls (accessed December 10, 2016).
International Telecommunication Union ITU (2016b) *ITU world telecommunication/ICT indicators (WTI) database 2016*, Geneva: International Telecommunication Union ITU.
Islam, M. S. and Grönlund, Å. (2011) "Bangladesh calling: farmers' technology use practices as a driver for development," *Information Technology for Development*, 17 (2): 95–111.
Kleine, D. (2010) "ICT4WHAT? Using the choice framework to operationalise the capability approach to development," *Journal for International Development*, 22 (5): 674–692. https://doi.org/10.1109/ICTD.2009.5426717.
Kopomaa, T. (2002) "Mobile phones, place-centred communication and neo-community," *Planning Theory and Practice*, 3 (2): 241–245. http://dx.doi.org/10.1080/14649350220150125.
Malpas, J. (2012) "The place of mobility: technology, connectivity, and individualization" in Wilken, R. and G. Goggin (eds) *Mobile technology and place*, New York: Routledge, 26–38.

Massey, D. (2005) *For space*, London: Sage.

Miller, D. (2009) "What is a mobile phone relationship?" in Alampay, E. (ed.) *Living the information society in Asia*, Ottawa: International Development Research Centre, 24–35.

Moores, S. (2004) "The doubling of place: electronic media, time-space arrangements and social relationships" in Couldry, N. and McCarthy, A. (eds) *Media space: place, scale and culture in a media age*, London: Routledge, 21–37.

Moores, S. (2012) *Media, place and mobility*, New York: Palgrave Macmillan.

Palackal, A., Nyaga Mbatia, P., Dzorgbo, D.-B., Duque, R. B., Ynalvez, M. A., and Shrum, W. M. (2011) "Are mobile phones changing social networks? A longitudinal study of core networks in Kerala," *New Media & Society*, 13 (3): 391–410.

Paragas, F. (2009) "Migrant workers and mobile phones: technological, temporal, and spatial simultaneity" in Ling, R. and Campbell, S. W. (eds) *The reconstruction of space and time: mobile communication practices*, New Brunswick: Transaction, 39–65.

Parvez, J., Islam, A., and Woodard, J. (2015) "Mobile financial services in Bangladesh: a survey of current services, regulations, and usage in select USAID projects," available online at: www.microlinks.org/sites/default/files/resource/files/MFSinBangladesh_ April2015.pdf (accessed August 6, 2016).

Pfaff, J. (2010) "Mobile phone geographies," *Geography Compass*, 4 (10): 1433–1447.

Rashid, A. T. (2011) "A qualitative exploration of mobile phone use by non-owners in urban Bangladesh," *Contemporary South Asia*, 19 (4): 395–408.

Reckwitz, A. (2002) "Toward a theory of social practices: a development in culturalist theorizing," *European Journal of Social Theory*, 5 (2): 243–263.

Schouten, P. (2012) "Theory talk #47: Jean Francois Bayart on globalization, subjectification, and the historicity of state formation," available online at: www.theory- talks. org/2012/02/theory-talk-47.html (accessed: October 16, 2017)

Shrum, W., Mbatia, P. N., Palackal, A., Dzorgbo, D.-B., Duque, R. B., and Ynalvez, M. A. (2011) "Mobile phones and core network growth in Kenya: strengthening weak ties," *Social Science Research*, 40 (2): 614–625.

Siddiqui, K., Ahmed, J., Siddique, K., Huq, S., Hossain, A., Nazimud-Doula, S., and Rezawana, N. (2010) *Social formation in Dhaka, 1985–2005: a longitudinal study of society in a third world megacity*, Aldershot: Ashgate.

Siddiqui, K., Gosh, A., Bhowmik, S. K., Siddiqi, S. A., Mitra, M., Kapuria, S., Ranjan, N., and Ahmed, J. (2004) *Megacity governance in South Asia: a comparative study*, Dhaka University Press: Dhaka.

Sterly, H. (2015) " 'Without a mobile phone, I suppose I had to go there': mobile communication and translocal social constellations in Bangladesh," *ASIEN*, 134 (1): 31–46.

Sterly, H. and Gerads, D. (2016) " 'Call me in the dorm': mobile communication and the shifting topographies of intimate relationships in Bangladesh," *Internationales Asienforum*, 47 (3–4): 273–269.

Tan, B. A. and Yeoh, B. S. (2011) "Translocal family relations amongst the Lahu in Northern Thailand" in Brickell, K. and Datta, A. (eds) *Translocal geographies: spaces, places, connections*, Burlington: Ashgate, 39–54.

UNDESA (2016) "Demographic Yearbook 2015, Table 1 Population, rate of increase, birth and death rates, surface area and density for the world, major areas and regions: selected years," available online at: http://unstats.un.org/unsd/demographic/products/ dyb/dyb2015/Table01.xls (accessed December 10, 2016).

United Nations (2012) *World urbanization prospects: the 2011 revision*, New York: United Nations.

Weber, M. (1922) *Wirtschaft und Gesellschaft, Grundriss der Sozialökonomik Vol. III*, Tübingen: J. C. B. Mohr.

Wellman, B. (2001) "Physical place and cyberplace: the rise of personalized networking," *International Journal of Urban & Regional Research*, 25 (2): 227–252.

Werlen, B. (2009) "Everyday regionalisations" in Kitchin, R. and Thrift, N. (eds) *International encyclopedia of human geography*, Amsterdam: Elsevier, 286–293.

Worldbank (2012) "Information and communications for development 2012: maximizing mobile," available online at: www.worldbank.org/ict/IC4D2012 (accessed May 30, 2014).

Zook, M., Dodge, M., Aoyama, Y., and Townsend, A. (2004) "New digital geographies: information, communication, and place" in Brunn, S. D., Cutter, S. L. and Harrington, J. W. (eds) *Geography and technology*, Dordrecht: Kluwer Academic, 155–176.

Part II

Subjectivities and identities

The subjective, perceptive, and affective engagement with the world, inherently co-constituted with the digital, is the focus of the following section. As the digital is "seeping" into everyday culture the "small worlds" of routines and life-world experiences reflect its shifts while the basic human modes of getting in touch with the world persist. Experiences and practices which are technically connected to global networks entail subjective and sometimes intimate moments of attachment to places; they can also promote a lack of orientation and even promote alienation, in the way we look at the world and being looked at, the way we move, navigate, or communicate and the ways in which we encounter others.

As Mike Duggan's contribution "The everyday reality of a digitalizing world" shows, meanings of space and place are integrated into the fabric of the lifeworld and, thus, digital culture. For example, manifold forms of "geoinformation" and "geovisualization" appear as an inherent part of digital culture's webs of meaning. From a reflexive perspective, everyday routines of interacting with digital devices stand for a remarkable transformation. Yet, in everyday life they happen mostly "on the fly" receiving little attention by the actors and are performed with a seemingly effortless ease.

The impression of a natural incorporation of digital technology into everyday life is widely shared—especially with regard to young people. The assumption of a "Generation Z" of digital natives often implies a notion of some kind of "natural" attunement of young people toward digital culture and, thus, a special competence in arranging or creating their individual lifeworlds. In their chapter "The emerging hegemony of cybernetic class n realities," Paul Montuoro and Margaret Robertson question this notion with regard to the role place and space play in processes of socialization and identity formation. Digital media and technology challenge (and change) the ways in which young people grow up as (not) being connected to or in touch with the "real world." As Montuoro and Robertson point out, identity building and identity crisis are deeply connected to the emergence of ubiquitous interfaces, social media, and the many forms of "distanciation" from the real world. Thus, a detachment from place takes hold of contemporary processes of identity formation.

A more affirmative view on the "realness" of digital culture's places and spaces is introduced in Pablo Abend's "From map reading to geobrowsing."

While others emphasize the great power with which geomedia platforms such as Google Earth perform the formation of worldviews, geopolitics, and common geographical imaginations, Abend focuses on the individual practices and subjective aspects of becoming engaged with the medium. New "action spaces" emerge between and around the screen and the user—offering involvement via affect, playful engagement, and bodily and cognitive practices exercised in front of and with the screen and the geobrowser's interface. In this sense, Abend shifts perspective from the analysis of power structures toward qualitative observations of the user's interaction with the screen or interface.

5 The everyday reality of a digitalizing world

Driving and geocaching

Mike Duggan

> We don't just use or admire technology; we live with it. Whether we are charmed by it or indifferent, technology is deeply embedded in our ordinary everyday experience.
>
> (McCarthy and Wright 2004: 2)

> I put the coffee down on the table, brought for enjoyment but mainly for the WIFI access that accompanies it. It's a particularly cold day in New York and I've had enough of standing outside fast food restaurants trying to pick up their WIFI signal all the while avoiding the offerings on the inside. The "free" WIFI offered at this particular coffee shop is akin to the elusive "free lunch". I feel inclined, and am certainly expected, to spend the $3.00 on a cappuccino in order to use the WIFI available. I'm clearly not the only one either; looking around this place it is alive with screen activity, which I'm sure is benefitting from the shop's connectivity.
>
> (Auto-ethnographic diary excerpt, New York, March 2015)

For those of us living in digitalizing worlds, these are familiar scenes. The ever-spreading use of digital technologies is transforming even the most banal instances of everyday life. There is little denying this, for digital technologies have become subsumed into almost all of our daily activities. Our mobilities, socialities, corporealities, and sensory experiences of the world have all been touched upon by digital technologies in one form or another, to a lesser or greater extent. This is to say that our everyday geographies have been thoroughly shaken up by the prevalence of digital technology. This chapter focuses primarily on the social and cultural implications of daily life interwoven with digital technology. The aim is to describe how digital technologies have become enfolded into the spatial flows of socio-cultural life in ways that appear at once, both seamless and disjointed.

In describing and examining the ways in which geographies are produced *by*, *through*, and *of* the digital, many geographers have noted the intrinsic ties between digital technologies and geographic practices (Ash et al. 2016). For geographers particularly interested in the spaces, places, and practices of everyday life, there has been little choice but to take note of these ties, for they have become inextricably linked in many cases. Much of everyday practice has become dependent or at least heavily reliant upon a wide variety of digital technologies. Thrift and French (2002: 309) argued over a decade ago that geographic research must take note of the "automatic production of space," for it was clear that software, automation, and computing systems were coming to produce the spaces of our daily practices. More recently, and in response to the continuing proliferation of software in everyday life, Kitchin and Dodge (2011) cemented this notion by giving explicit testimony to the fact that the production of (social) space is heavily dependent on the digital code(s) that now produce it. They made the convincing argument that the everyday practices of modern metropolitan societies increasingly unfold in either a configuration of code/space or coded spaces; that is to say, spaces which occur "when software and the spatiality of everyday life become mutually constituted" (ibid.: 16), or spaces in which "software makes a difference to the transduction of spatiality but the relationship between code and space is not mutually constituted" (ibid.: 18). The key point of their thesis was to clarify the relationship between digital technology and the production of spatial practices.

Using these broadly accepted arguments, scholars have sought to highlight the specifics of code/space and coded space by focusing their interests on the impacts of digital technologies across a broad range of topics, across many disciplines. Geographic research has detailed how digital technologies broadly and specifically effect economic and political place-making and urban sociospatial practices (see Elwood and Leszczynski 2013; Graham et al. 2013; Leszczynski 2012; Zook and Graham 2007). Media and communication scholars have been particularly forthright in researching the impact of mobile technologies on the sociality, locality, and the phenomenology of place (see Evans 2015; Frith 2015; Gordan and de Souza e Silva 2011; Wilken and Goggin 2012), and a growing number of anthropologists working under the umbrella of "Digital Anthropology" have produced pertinent descriptions about the role that digital technologies play in everyday life (see Hjorth and Richardson 2014; Horst and Miller 2012; Ito et al. 2005; Miller and Sinanan 2014; Pink 2012; Pink et al. 2015; Tacchi 2012). Threading through all of these writings is the notion that space, as conceived, perceived, and lived (after Lefebvre 1991), and place, as the processual experience of space (after Cresswell 2004; Massey 2005), are socio-technical formations increasingly co-constituted by entanglements of digital technology and socio-spatial practice. The underlying argument implied in many of these writings is the suggestion that space and place are becoming increasingly *augmented, hybridized, layered,* or *mediated* with the information made accessible by digital devices. As noted by Graham et al. (2013), access to such a dynamic set of digital information at our fingertips has the capacity to produce space and experiences of place in novel ways.

In this chapter I extend and critically explore these arguments by examining the minutia of everyday mapping practice. Whereas existing geographical work has been key to understanding the impacts of digital technology on society more broadly, to date very little research has focused on the intricacies of everyday life. Using excerpts taken from ethnographically led field work, aimed at capturing the state of contemporary mapping practices in London and the Southeast of England (UK), I wish to address this gap in the research by highlighting the intricacies of *socio-technical situations* (see Ito et al. 2005). I hope to provide a rarely voiced perspective on how digital technologies augment our experiences of place. In doing so, I will show how digital technologies have the capacity to produce novel socio-spatial experiences of the world at the level of everyday practice.

I will develop two interconnected notions. First, I wish to expand on Kitchin and Dodge's (2011) thesis of code/space and coded space and its relation to a processual understanding of place, for rarely is their influential theory collated with those concerning place. In defining a processual understanding of place as an ever-unfolding constellation of processes, practices, materials, histories, possible futures, and experiences, rather than simply location or locale (after Massey 2005), I seek to examine the ways in which code/space and coded space can affect such constellations in everyday practice. In doing so, I pay particular attention to how a *sense of place*—the key experiential and perhaps most porous element of place's constitution (see Agnew 1987; Relph 1976; Tuan 1977)—can be affected in configurations of code/space and coded space. Laying out two descriptive accounts of driving and navigational practice, I detail the ways in which experiences of place may become inextricably linked to the use of smartphone, digital radio, and digital mapping technologies. Offering an alternative perspective to scholars certain that the digitalizing world deprives us of a sense of place (see Augé 1995; Meyrowitz 1985), I argue that a sense of place is increasingly constituted through the socio-technical coming-together of digital technology, culture, and practice. Following Coyne's (2010) assertion that place is increasingly *tuned* by the digital devices available, I describe this constitution in terms of how people may use digital technologies to *tune* and *re-tune* everyday experiences of place.

In the second section of the chapter I shed light on a further example, which highlights the social experiences made possible by socio-technical situations. This section extends the previous assertions out of individual experiences of place and into collective experiences of place. Using the example of Geocaching—a treasure hunting game made possible by location-aware technology—I suggest that the socio-technical constitution of digital technology and culture has the capacity to produce novel socio-spatial relations. In other words, I use this example to describe how digital technologies can foster novel forms of culture, sociality, and entertainment in everyday practice.

Ultimately, this chapter intends to describe and explore how digital technology and culture are now mutually constituted in everyday practices and consequently in everyday experiences of place (see Castells 2000; Horst and Miller 2012; Kitchin and Dodge 2011; Moores 2012; Pink et al. 2015; Postill 2011).

Digital technology and a sense of place

Unsurprisingly, the sleek picture of a life with technology painted across advertisements is rarely forthcoming when examining the practices of everyday life. The reality of these entanglements is predictably messy (see Shove 2007). Indeed, my ethnographic study into how people used digital maps in their everyday lives rarely presented the use of these technologies as straightforward and neutral processes of application. The overwhelming finding from this study was that the socio-technical constitution of everyday digital mapping practice is a context dependent process. In the following, I use my experiences with Tom (aged 26) and Sally (aged 28), two of the research participants taken from this study, to demonstrate this point. In providing descriptions of Tom and Sally's driving practices, taken from accompanying each of them on two drives lasting for 2 hours, I show how digital mapping technologies can be enfolded—quite peculiarly and specifically at times—into everyday experiences of driving and place-making. I compare and contrast these insights using ethnographic descriptions and analysis in order to highlight the different ways in which digital technologies can become embedded in everyday driving practices, which ultimately produce differing experiences of place. By doing so, I argue that processes of place-making in the digital age are constituted by complex and context dependent configurations of technology, practice, and embodiment.

Tom: tuning a (mobile) sense of place

> We sat stationary in the car for a few minutes outside the station while he typed our destination into GoogleMaps, his go-to mobile application for navigation. He then placed the phone, screen-up, on the car's central console rather than on the inside of the windshield, as is common for many people. The bracket he did have had broken months ago, and so he was forced to fashion a suitable place to see the phone from the limited apparatus he had available. After inputting his destination, he went to the settings menu, turned off the default voice navigation function and then switched on the radio.

Both of these acts were considered essential for creating the kind of experiences he wanted to have while driving. Driving, for Tom, was a meditative exercise of sorts. He claims that he was aware of his actions but uses the practice as a way to switch off from almost everything else, including paying little attention to the places we were driving through. He explained that the motions and kinesthetic sensations of driving created pleasing rhythms in which he could relax and not think about the stresses of life and work. While the action of intermittently checking his mobile phone for directions amid the ambience of afternoon radio was not seen by him to break these rhythms (it had become a part of them) the interjection of verbally assisted directions was. The voice was "distracting and annoying," he said. It took him out of his zone. In this sense,

Tom was able to partly curate his driving experience using the digital technologies available to him.

These insights highlight one such instance in which the spatial practices of driving are produced, in part, by the digital technologies involved. In this case, a smartphone loaded with Google Maps and the car's radio were significant in the constitution of Tom's driving practices. Without them, driving would be different for Tom, which suggests that digital technologies had come to affect the experiential element of his driving practices. He purposely appropriated these technologies to curate an experience of driving which suited his needs. He had, in other words, intentionally *tuned* his use of technology (see Coyne 2010) in order to try and produce a certain sense of place, to make sense of his (mobile) place. Place is not simply a location or locale in this regard; it is neither the locale of the car nor his global, regional, or local position. Rather, it is an event that unfolds as a constellation of trajectories that include multiple processes, practices, histories, possible futures, and experiences (Massey 2005). In his *tuning* of technology, it could be said that Tom has attempted to make sense of and control these ongoing flows of place. However futile these adjustments may appear when considering the rapidly changing practices of driving, it is Tom's way of bringing some stability to these processes. Moreover, in following Miller and Sinanan's (2012) theory of attainment, it could also be said that Tom had made real, through his practices of tuning, the latent capacity of these technologies to suit his particular desire to control the unfolding constellation of place. In other words, he was appropriating technology in a way that suited his desire to attain the driving zone he wanted.

Tom has not simply appropriated the technology in the ways that are prescribed by its makers, but rather he has intertwined them with his own cultures of practice, many of which are unique to him. At the very least, the example of Tom highlights the dialectical relationship between technology and culture, and it rejects any notion that experiences of place can be technologically determined.

Sally: tuning an embodied sense of place

> Secured by a bracket attached to the windshield, Sally tapped our destination into Google Maps. Her phone has become inextricably linked to her practices of driving. It is often consulted, tapped, swiped, adjusted. Her attention is divided between the phone and the road.

Google Maps was not always used specifically as a turn-by-turn navigational aid; Sally would often switch between a turn-by-turn view to a view that displayed the entire route-map. She would frequently reach out to switch between views on our trips together. This way, she could switch between a map that gave her the reassurance that she was traveling in the right direction and a map that gave her the immediate directions she needed, something which she considered entirely necessary. The turn-by-turn view was used to accommodate the more immediate need to know which way to go, and the zoomed-out view was used to assess where she was in relation to where she was going and also as a tool to plan where she could take a suitable break. Taking breaks, she told me, had become an important part of Sally's driving practices in recent years. The niggling pain of an ongoing spinal injury often forced her to do so, especially when she was required to travel over long distances. Google Maps has by no means solved this issue, but it had helped her to manage it on a day-to-day basis. It provided her with real-time travel information in ways that paper maps could not do, something which she used to judge how traffic might affect her journey time and access to service stations and rest stops. When commenting on this way of use, Sally acknowledged its peculiarity but claimed that having tried the "proper" way of using the software, she found this way much more suited to her style of driving and navigation.

Similarly to Tom, Sally had appropriated the technology, through practice, to curate the driving environment she wished to have. She had used digital mapping technology to curate a socio-technical situation that served her corporeal need to take regular breaks from driving. Tom's experiences of socio-technical situations appear far more bound up in the atmospheres of driving; whereas, Sally's experiences have become tied to the sensory and embodied experiences of her lifeworld (see Pink and Mackley 2013). In a processual understanding of place, a sense of place is fundamentally unstable (Anderson 2012). The example of Sally highlights how the instability of place can be intertwined with the digital technologies available at hand. Throughout her driving practices, Sally frequently tuned and re-tuned her sense of place using the digital mapping technology available to her. Indeed, it could be said that she used the dynamic and wider connectedness of this technology to accommodate novel ways of dealing with the unpredictable need she had to regularly stop and take a rest from driving. This shows one way in which socio-technical situations may be different to those produced without digital technology.

For Coyne (2010: 16), "the tuning of place is a set of practices by which people use devices, willfully or unwittingly, to influence their interactions with one another in places." Using these examples, I suggest that both Tom and Sally actively tuned their sense of place based on the technologies they had on hand. *Tuning place*, I suggest, is not simply determined by how devices influence interactions with *who* is present or absent, as Coyne suggests, but also by how devices influence interactions with practice itself, for practice is fundamental to the experience of place (Shove et al. 2012).

Following Kitchin and Dodge (2011), Tom and Sally's experiences could be said to be unfolding as differing constitutions of coded space, rather than as a product of code/space. The reason being that their practices of driving and navigation would remain possible if Tom and Sally did not make use of smartphones, radios, and digital mapping technology. The drivers themselves made conscious decisions to appropriate the space(s) of the car by incorporating these technologies into their respective cultures of driving practice, and their individual preferences highlighted how they had influence over how the space was appropriated by technology. Nevertheless, for Sally and Tom, the experiences of driving rarely, if ever, unfolded in spaces absent of digital technology. These included both the technologies that I have described and the technologies that I have not. There is clearly the car itself, GPS satellites, and traffic monitoring technologies among many others to consider when examining the assemblages of such socio-technical situations. In this case it could therefore be argued that their experiences were mediated by multiple technologies and are indeed code/ space. I suggest that these experiences of place would not be possible without digital mapping technology because the technologies themselves have become instrumental in the constitution of Tom and Sally's driving experiences. Following Leszczynski (2015), their experiences could be described as *always-already* mediated by the many ways in which digital technology now co-constitutes driving practice.

In the same way that a theory of code/space edicts the constitution of code and the production of space, I suggest that certain experiences of place, when understood as a processual concept, are only made possible by the co-constitution of technology and culture in practice. They are *codified experiences of place*, or to use Kitchin and Dodge's terminology, experiences of *code/place*. When comparing Tom and Sally's practices it is clear that there can be many nuances to the ways in which such constitutions are experienced in everyday life. As these case studies highlight, codified experiences of place unfold in the unique milieu of one's social, cultural, and embodied lifeworld.

Digital technology and socio-spatial relations

So far in this chapter I have sought to describe and explore the dialectical relationship between digital technology and culture in everyday practice. I have used examples to suggest that it is the cultures of practice that mediate the relationship between digital technology and what it means to experience place in the contemporary world. I have suggested that place, as an inherently processual concept, is increasingly experienced as a *socio-technical situation* in which we *tune* and are *tuned* by the technologies we have available on hand. In this second section I develop these ideas further with regards to how digital mapping technologies can be used to produce novel forms of spatial practice, which I argue come together with an assemblage of non-digital socio-material practices to constitute a sense of place. I describe the everyday practices of Geocaching—a treasure hunting game and community made possible by location-aware GPS technology—in order to give an example of how these socio-technical and socio-material relations may unfold in everyday mapping practice.

Geocaching is defined as a "real-world, outdoor treasure hunting game using GPS-enabled devices [whereby] participants navigate to a specific set of GPS coordinates and then attempt to find the geocache (container) hidden at that location" (geocaching.com 2016). Since its inception in 2000, it has become a popular location-based recreational activity that is played around the world. In order to play, participants are encouraged to find and hide a variety of cache types. In order for a found cache to count, a player must log it both in the cache's logbook and also on the geocaching.com website or via a smartphone application.

The premise of Geocaching is not particularly new; it extends from earlier non-digital community-based scavenger hunting games. A hobby that it shares many features with is Letterboxing; an outdoor hide and seek activity that brings together problem solving, orienteering, hiking, and practices of exchange that dates back to the mid-nineteenth century (see Hall 2003). This significantly predates the existence of Geocaching, which took off in the early 2000s. What is novel about Geocaching is the ways in which the practice is intertwined with digital technologies. Indeed, for players to be fully engaged in Geocaching they must use a number of digital devices and software. For instance, to hide or seek a cache—the basic premise of Geocaching—players must make use of GPS technology. Moreover, to log caches on the official system—something that is very important for committed players—players must use a computing device connected to the internet. In this regard, the practice could be said to unfold as a constitution of code/space (Kitchin and Dodge 2011). This is to say that the (social) production of Geocaching spaces is reliant upon the technologies which have come to produce them. Further to this and with reference to the previous section, it could be said that the experiences of Geocaching spaces are also not possible without the digital technologies implicit in their workings. Ultimately, I suggest that the practices and experiences of Geocaching are constituted by the coming-together of digital technologies—primarily digital maps, GPS, and internet technologies—and everyday cultures of practice.

In the following I provide an example of how novel forms of socio-spatial relations and experiences of Geocaching unfolded for a group of caches primarily based in South London, UK. The cases below are taken from an ethnographic study of this group. Once again I use excerpts and analysis taken from this ethnography in order to highlight the unpredictable and messy geographies of everyday practices as they become intertwined with digital technology. In the following extract I describe how Geocaching can produce novel practices of exchange within a community group.

Accompanying many caches are Trackable Bugs (TBs). These are usually small toy-like objects collected by cachers in order to be exchanged with others at community meetings. By doing so, cachers add another layer to the game which is to collect, exchange, and record as many TBs as they can. During one such meeting, I made note of how these interactions took place.

> The swapping came fairly early on, which was unnerving for me at first. People began to empty their bags of what appeared to be trinkets of some kind or another and began to exchange them with one another. I was perplexed, stumped even, as Debra swapped a miniature London red bus with Simon's bathtub duck.

For many around the table, this was a good opportunity to shift TBs they had been hoarding and add a few more points to their total, which is something everyone was pleased to do. While the collecting element was not necessarily the strongest motivating factor for this group of cachers, it was clearly important for those involved, not only for points collected, but also for the kind of social adhesive it produced. These practices of exchange brought people together using a commonality that they were all familiar and comfortable with. Purposefully initiated at the beginning of meetings, these practices effectively broke-the-ice on the socialites of the evening. It could be said that these practices facilitated a *geography of enthusiasm* (see Geoghegan 2013).

Despite the lack of technology used or even talked about at these meetings, such practices could not have unfolded independently of digital technology. Indeed, as Pink et al. (2015: 57) argue, "researching digital media practices often actually means researching the relationship between digital media and other things and processes, and considering how the practices through which these are played out become blurred." I took a non-media-centric approach (see Moores 2012; Morley 2007) to exploring the use of digital media in everyday practices of Geocaching, and as such, it was found that digital technologies lay at the foundations to many socio-spatial practices. This was useful in uncovering the often hidden links tying digital technology to Geocaching. Throughout these exchange meetings, there was often little digital technology to see, let alone seen to be used in the exchanges themselves. Most of the group preferred to use pens, pencils, and note pads to keep a record of exchanges rather than make notes using phones, tablets, or laptops. Among the swapping, there was also a good deal of recording going on. Some would be writing down the TB information in pocket-sized notebooks in order to log the information on "geocaching.com" at a later date. Additionally, a notebook was passed around to everyone at the meet-ups to record attendance. These actions were an easy and accessible way to quickly jot down cache and event information to be logged at a later date. Often going unnoticed as technologies, such tools clearly still continue to co-constitute the digitalizing world. Indeed, while digital technologies are increasingly coming to produce novel socio-technical spatial practices, rarely, if ever, do they emerge in a vacuum that is absent of other, older forms of technology. Socio-technical situations speak to all forms of technical interaction and should not necessarily be predisposed as digital. A messy assemblage of digital and analog technology now constitutes much of what we do. This is something to consider in many of our everyday practices and not something unique to Geocaching. We live in a material world in which digital technologies and their affordances are merely another layer, rather than a separate realm of existence (see Graham 2010). In the case of Geocaching, social practices are produced by socio-technical situations, but they are also produced between players and non-digital things throughout collective caching expeditions, within exchange events, and over pints in the pub. In this manner, Geocaching is a great example of where the digital meets the analog in everyday practice.

While Geocaching is a treasure hunt, it is often so much more. As well as simply being fun, it is often referred to as a good form of exercise, an opportunity to visit places off the tourist trail, a challenge, a good source of friendly competition, an adventure, and a good way to meet like-minded people. Many people have incorporated the practice into multiple areas of their life. One of the participants, Tara, went as far to suggest that it can "take over your life!" when referring to people who channel large parts of their social and leisure time into Geocaching-related activities. Two others, Robert and Louise, were the first to admit this, noting how most of their recent holidays were to places rich in unfound caches. The previous summer both had attended a week-long caching event on the Isle of Man; the purpose being to follow an extended multi-cache trail and socialize with friends they had made while Geocaching over the years.

The spaces and experiences of Geocaching were produced through a broad constitution of processes, practices, and technologies, be they digital or analog. Practices of exchange and social gathering are clearly more than just a product of digital technology. Nevertheless, it would be difficult to ignore the vital part that digital technologies perform in the production of Geocaching practices and experiences. For instance, in order to come together and exchange TBs, players must first use GPS devices to find caches that contain TBs. Second, in order for members to find out about exchange groups—where they are, who is going, and what kind of event it is—they are required to consult the internet for information, either through official Geocaching networks, social networking platforms, or via one-to-one messaging services. Third, in order for members to log TBs on their official account pages they are required to use a computing device connected to the internet. Without the coming-together of these digitally led practices such exchanges and the kinds of sociality that they produced would not have been configured in the same way. This, I suggest, is how digitally led practices can produce novel forms of socio-technical spatial experience. Following Massey (2005: 41), the events of place can be thought of as a "constellation of processes." Using the example of Geocaching, I argue that digital technologies now have a significant, but not necessary exclusive, role to play in the constellations of place and everyday practice.

Conclusion

Digital technologies are increasingly intertwined with the geographies of everyday practice. In many cases, they are already inextricably bound. By describing, analyzing, and comparing the differences in how everyday instances of driving, navigation, and leisure practices may be bound up in this constitution, this chapter sought to highlight the daily, often mundane, and idiosyncratic realities of those living in a digitalizing world. In the first section I used the example of Tom and Sally, two research participants, to show how drivers curated their use of Google Maps to *tune* what I called their mobile and embodied sense of driving places. In the second section I outlined how an assemblage of digital and non-digital material practices constituted the socio-technical spaces of a

Geocaching community. The purpose of this was to show how deeply embedded digital technologies have become in everyday practices.

This chapter provides an empirical perspective on the more theoretical work being done on this topic, particularly within the discipline of Human Geography. I have suggested that the unfolding constellation of place and social practice are increasingly experienced as *socio-technical situations* (see Ito et al. 2005) in which we are able to *tune* place and be *tuned* by the technologies in place (Coyne 2010). In laying out these assertions, I have brought the notion that digital technologies increasingly produce the spaces of our everyday lives (Kitchin and Dodge 2011; Thrift and French 2002) into focus with the conceptual debates around the processuality of place, arguing that the coded production of space must take into account the everyday experiences of this constitution. Ultimately, this chapter has offered a fresh disagreement with the notion that a digitalizing world deprives us of a sense of place (see Augé 1995; Meyrowitz 1985). As I have described it, digital technologies undoubtedly complicate, augment, and affect our experiences of place, but they do not necessarily detract from our sense of place.

Acknowledgments

This chapter has been written with the support of the Engineering and Physical Sciences Research Council (EPSRC) and the Ordnance Survey (ICASE award).

References

Agnew, J. (1987) *Place and politics: the geographical mediation of state and society*, London: Allen & Unwin.

Anderson, J. (2012) "Relational places: the surfed wave as assemblage and convergence," *Environment and Planning D: Society and Space*, 30 (4): 570–587. https://doi.org/10.1068/d17910.

Ash, J., Kitchin, R., and Leszczynski, A. (2016) "Digital turn, digital geographies?," *Progress in Human Geography*, Online first. https://doi.org/10.1177/0309132516664800.

Augé, M. (1995) *Non-places: introduction to an anthropology of supermodernity*, London: Verso.

Castells, M. (2000) *The rise of the network society, The information age: economy, society, and culture Vol. 1*, Oxford: Wiley.

Coyne, R. (2010) *The tuning of place: sociable spaces and pervasive digital media*, Cambridge, MA: MIT Press.

Cresswell, T. (2004) *Place: a short introduction*, Oxford: Wiley-Blackwell.

Elwood, S. and Leszczynski, A. (2013) "New spatial media, new knowledge politics," *Transactions of The Institute of British Geographers*, 38 (4): 544–559. https://doi.org/10.1111/j.1475-5661.2012.00543.x.

Evans, L. (2015) *Locative social media: place in the digital age*, London: Palgrave Macmillan.

Frith, J. (2015) *Smartphones as locative media*, London: Wiley.

Geocaching.com (2016) "What is geocaching?," available online at: www.geocaching.com/guide/ (accessed December 12, 2016).

Geoghegan, H. (2013) "Emotional geographies of enthusiasm: belonging to the telecommunications heritage group," *Area*, 45 (1): 40–46.

Gordan, E. and de Souza e Silva, A. (2011) *Net-locality: why location matters in a networked world*, Oxford: Wiley-Blackwell.

Graham, M. (2010) "Neogeography and the palimpsests of place: web 2.0 and the construction of a virtual earth," *Tijdschrift voor economische en sociale geografie*, 101 (4): 422–36.

Graham, M., Zook, M., and Boulton, A. (2013) "Augmented reality in urban places: contested content and the duplicity of code," *Transactions of the Institute of British Geographers*, 38 (3): 464–479. https://doi.org/10.1111/j.1475-5661.2012.00539.x.

Hall, R. (2003) *The letterboxer's companion*, Guilford: The Globe Pequot Press.

Hjorth, L. and Richardson, I. (2014) *Gaming in social, locative and mobile media*, London: Palgrave Macmillan.

Horst, H. and Miller, D. (2012) *Digital anthropology*, London: Berg.

Ito, M., Matsuda, M., and Okabe, D. (2005) *Personal, portable, pedestrian: mobile phones in Japanese life*, Cambridge MA: MIT Press.

Kitchin, R. and Dodge, M. (2011) *Code/space*, Cambridge MA: MIT Press.

Lefebvre, H. (1991) *The production of space*, Oxford: Wiley-Blackwell.

Leszczynski, A. (2012) "Situating the geoweb in political economy," *Progress in Human Geography*, 36 (1): 72–89.

Leszczynski, A. (2015) "Spatial media/tion," *Progress in Human Geography*, 39 (6): 729–751.

Massey, D. (2005) *For space*. London: Sage.

McCarthy, J. and Wright, P. (2004) *Technology as experience*, Cambridge MA: MIT Press.

Meyrowitz, J. (1985) *No sense of place: the impact of electronic media on social behavior*, Oxford: Oxford University Press.

Miller, D. and Sinanan, J. (2012) "Webcam and the theory of attainment," Working paper for the EASA media anthropology network's 41st e-Seminar, October 9–23, 2012, available online at: www.media-anthropology.net/file/miller_sinanan_webcam.pdf (accessed July 26, 2017).

Miller, D. and Sinanan, J. (2014) *Webcam*, London: Polity Press.

Moores, S. (2012) *Media, place and mobility*, London: Palgrave Macmillan.

Morley, D. (2007) *Media, modernity and technology: the geography of the new*, London: Routledge.

Pink, S. (2012) *Situating everyday life*, London: Sage.

Pink, S., Horst, H., Postill, J., Hjorth, L., Lewis, T., and Tacchi, J. (2015) *Digital ethnography: principles and practice*, London: Sage.

Pink, S. and Mackley, S. L. (2013) "Saturated and situated: expanding the meaning of media in the routines of everyday life," *Media, Culture and Society*, 35 (6): 677–691.

Postill, J. (2011) *Localizing the internet*, Oxford: Berghahn.

Relph, E. (1976) *Place and placelessness*, London: Pion Press.

Shove, E. (2007) *The design of everyday life*, New York: Berg.

Shove, E., Pantzar, M., and Watson, M. (2012) *The dynamics of social practice*, London: Sage.

Tacchi, J. (2012) "Open content creation: the issues of voice and the challenges of listening," *New Media and Society*, 14 (4): 652–668.

Thrift, N. and French, R. (2002) "The automatic production of space," *Transactions of the institute of British geographers*, 27 (3): 309–335.

Tuan, Y. (1977) *Space and place: the perspective of experience*, Minneapolis: University of Minnesota Press.

Wilken, R. and Goggin, G. (2012) *Mobile technology and place*, London: Routledge.

Zook, M. and Graham, M. (2007) "Mapping digiplace: geocoded internet data and the representation of place," *Environment and Planning B: Planning and Design*, 34 (3): 466–482.

6 The emerging hegemony of cybernetic class *n* realities

The non-place of Generation Z

Paul Montuoro and Margaret Robertson

No longer to travel, except on the spot. No longer to stretch ourselves, to spread ourselves thin in the passing distraction of the physical journey, but just to relax here and now, in the inertia of immobility regained. Social quietism leads our societies to wrap themselves in the shroud of interior comfort, the bliss of a reverse vitality in which lack of action becomes the height of passion. A society of hardened lounge lizards, where everyone hopes not to die or suffer, as western masochism has been said to want, but to be dead.

(Virilio 2000a: 62f.)

Traveling without moving

The friction of everyday life is changing, especially for young people. Individuals who were born between 1995 and 2010 (Generation Z), grew up *with* the internet (Seemiller and Grace 2016). Now, one of the challenges for researchers is to understand the implications of an increasingly technological society on Generation Z, the generation that never knew the world without the internet. Central to this chapter is the contention that Generation Z's interactions with "the real" have become increasingly mediated by technology. "The real" in this chapter refers to everything in the world that is experienced directly by the individual (Debord 2008). Furthermore, this chapter contends that through this mediation of the lived experience, or what Virilio (2000b: 3) calls "cybernetics," the phenomenology of perception has fundamentally transformed; young people no longer strictly inhabit the real but co-inhabit the expansive non-places of substitute realities almost entirely represented by pixels on a screen. The implications for the identity of individuals in Generation Z may be momentous, especially if identity is even partially a function of the physical environment (Erikson 1970).

Remembering the past helps to make sense of the enormity of the cybernetic realm. For example, we know from artists' images that the past was *different*. While the artist's imagination provides a subjective view to life in former times, we tend to accept these images as representative of some degree of truth. Like the depictions of bustling Flemish villages in a Bruegel painting (e.g., Bruegel 1559), or the social collaboration in one of Guys' *Winter scenes* of downtown Brooklyn (e.g., Guy c.1819–1820), the *Gemeinschaft* (or "community") of the erstwhile village evokes imagery of social relationships based on familiar bonds where everyone knows their role and responsibilities as well as everyone else's. Nowadays, places such as villages, towns, and even cities are increasingly being replaced by cybernetics as the location of formerly communal lived experiences. This supports Archer's (2013: 131) hypothesis that the sense of community experienced by former generations is indeed "giving way to something else."

An exploration of the lived experiences of Generation Z elicits thoughts of a vast and constantly metamorphosizing range of cybernetic mediators. For this generation, growing up *is* a digital experience. Wireless data streams imperceptibly to digital devices—text, images, and video (all of the imposters of the lived experience) appear apparently from *nowhere*. These devices provide access to a world of otherness *without the other*. Baudrillard wryly commented on this strange existence, in which space is a mere fiction and one's immediate physical surroundings are immaterial to the lived experience:

> Today things travel faster from Paris to Rio than from the hand to the brain. You have only to think of a friend at the other end of the world to be connected to him in real time. There is something inhuman in this instantaneity. I didn't necessarily want to be immediately in contact with him. I just wanted, perhaps, to toy with the idea of seeing him.
>
> (Baudrillard 2006: 26)

Augé (1995: 80ff.) explains that here, there is no development of "anthropological place," which refers to the evolving social history of a physical place that is relational, historical, and concerned with identity. Instead, there is a solitary passage through roughly equivalent "non-places," where there is no social integration of the past and where the individual simply "exists" anonymously in a sterile permanent present. For Augé (ibid.: 83), the individual in the non-place becomes, "no more than what he does or experiences in the role of passenger," leading to a "gentle form of possession." This leads to a temporary identity loss, briefly relieving the individual of worldly obligations, and in turn, allowing the possibility for the pleasure of role-playing. However, Augé (ibid.: 83) warns that here, all the individual confronts is a "pretty strange" image of him- or herself, and adds that, "the space of non-places creates neither singular identity nor relations only solitude and similitude."

Although we agree with Augé's perspective, we contend that the growth of the cybernetic realm, with its increasing levels of digital immersion, will result in more far-reaching changes to the identity of individuals in Generation Z, leading to a disruptive and lasting form of possession. Indeed, we contend that the advent of technology such as "neural dust," which is designed to treat diseases by inhabiting the space of the body *itself* (Seo et al. 2016: 529), may distance Generation Z from the real so dramatically that it will fundamentally change the course of human identity. Referring extensively to the work of the cultural theorist, Paul Virilio, this chapter discusses four stages of technological speed progression, which we contend are gradually leading to the disappearance of the human subject from the real, displacing the body as the center of lived experience in the world. Each discussion focuses on how the stage of progression appears to be altering identity formation in Generation Z.

Stage 1: the development of speed

It can be said that the preliminary stage of the development of speed in human history began to end sometime during the two decades of World War II. This stage began with persistence hunting by early *Homo*, a practice that involves chasing animals to exhaustion during the hottest part of the day (Liebenberg 2006). This stage progressed with the development of wooden projectile weapons by *Homo erectus* during the Lower Paleolithic (~400 Ka) in Germany (Thieme 2000), and stone-tipped projectile weapons by *Homo sapiens* during the Upper Paleolithic in the Levant and Europe, and during the late Stone Age (40–50 Ka) in Africa (Shea 2006).

This preliminary stage of ever-increasing (physical) speed reached a zenith and an endpoint in the early twentieth century with the proliferation of the internal combustion engine in commercial vehicles, especially the automobile (Virilio 2005). Indeed, whereas projectile weapons merely distanced early hunters from their prey, the automobile distances its occupants from interacting with the very environment itself. Urry (2007: 127) explains that, "although auto mobility is a system of mobility, it necessitates minimal movement once strapped into the driving seat." The result is that, enclosed in the vehicle's cabin, the occupant, who inhabits another place, no longer experiences the world directly, which leads to a disembodied experience. Ihde (1974: 272) describes this alienating effect of the automobile:

> The expert driver when parallel parking needs very little by way of visual clues to back himself in the small place – he "feels" the very extension of himself through the car as the car becomes a symbolic extension of his own embodiedness.

Furthermore, Virilio (2005) argues that increasing speed has a progressively transformative effect on the lived experience. For example, from the perspective of an occupant in a speeding automobile, a stationary object such as a tree appears to be in violent motion. What appears to the occupant is a distorted by-product of speed, forcing them to experience place in a fragmented manner. Speed beyond the capacity of human locomotion, "perverts the illusory order of normal perception, the order of arrival of information" (Virilio 2009: 110). Therefore, reality itself becomes "the first victim of speed ... what happens more and more quickly is perceived less and less distinctly" (Virilio 2005: 116). For Virilio, speed beyond the capacity of human locomotion leads to the *substitution* of "class I reality" with "class *n* realities" (Armitage 2000a: 43). Here, the term class *n* realities refers to any number of perceived realities existing simultaneously. Additionally, although Virilio explains that there have been different categories of socially constructed class *n* realities since the Neolithic age, the automobile represented a significant new class of *physically constructed* reality. Indeed, it has allowed every generation since the two decades of World War II, including Generation Z, to traverse large swathes of the landscape without *directly experiencing* them, at speeds too high for the individual to fully absorb what is given to consciousness.

Nevertheless, Matless (2016: 96) describes the primacy of place in the formation of identity during the proliferation of the automobile, when private ownership in England rose from 78,000 in 1918 to over two million in 1939. Matless' book, *Landscape and Englishness*, considers the period from 1918 to 1950, describing the significant role of the "geographical self" in England during this period. The term geographical self refers to the internalization of wider spatial relations in the formation of self-identity (for a review, see Matless 1997). Matless (2016: 29) explains that what space, and specifically the English landscape, "is" or "means" during this period, can "always be subsumed in the question of how it works; as a vehicle of social and self-identity, as a sight for claiming of a cultural authority." Whether place continues to play this critical role in the future, especially for Generation Z, may be dependent on the degree to which speed of light technologies *disconnect* the individual from the real.

Stage 2: the motionless motor

The increased use of cybernetic technologies in the second half of the 20th century marked the beginning of the submergence of physical geographical space as the location of human interaction with the real. Drawing on an interview with Virilio, Armitage remarks:

> Sure, transportation has been constantly speeded up to, but, today, the major development is the … quest for the attainment of real-time. Information transmission is thus no longer concerned with the bringing about of a relative gain in velocity, as was the case with railway transport compared to horsepower, or jet aircraft compared to trains, but about the absolute velocity of electromagnetic waves.
>
> (Armitage 2000a: 36)

Virilio (1991: 18) claims this has led to the "exhaustion of [the] physical." He suggests that we now exist in a *post-physical* condition, whereby traditional topographical and architectural constraints have shifted as a result of advances in technology. Indeed, with the increasing adoption of speed of light technologies, "it is no longer necessary" for anyone to "make any journey" since "one has already arrived" (Virilio and Lotringer 2008: 64). It is entirely possible that this is leading to a kind of Foucauldian incarceration à la Howard Hughes, whereby cybernetics assumes an all-encompassing, borderless carceral function, which leads to a de-realization of everything that was once directly lived (Armitage 2000a). Here the friction of the real is replaced by the frictionless lived experience associated with cybernetic class *n* realities.

This raises the need to redefine Husserl's (1970: 69) very notion of the *Lebenswelt* (lifeworld), which he defines as the world of pure experience that allows us to perceive nature, from which we can develop our cultural form. Briefly, Husserl's lifeworld is the unmediated perception of reality (for example, seeing a tree), or the material-mediated perception of reality (for example, feeling a tree through a glove). On the other hand, although the "cybernetic lifeworld" is not exactly the alien world described by Husserl, it is the perception of that which is fundamentally *non-local*. Here, the rapidity of cybernetics has obliterated the limitations of distance and space for the human subject. Virilio describes the impact of this phenomenology without the real:

> Thanks to satellites, the cathode-ray window brings to each viewer the light of another day and the presence of the antipodal place. If space is that which keeps everything from occupying the same place, this abrupt confinement brings everything precisely to that "place," that location that has no location. The exhaustion of physical, or natural relief and of temporal distances telescopes all localisation and all position. As with live televised events, the places have become interchangeable at will.
>
> (Virilio 1991: 17f.)

These technologies do not represent a vanishing point for the individual, but rather, a new kind of "uni-spatial existence." As opposed to the automobile, which led to a kind of autonomous movement for the human subject, cybernetics marked the beginning of a remote movement for the human subject. Here, the individual remains located in time and space, but what is given to consciousness is a strange *re-presentation*, or complete simulation, of reality. For the individual, the proliferation of the cybernetic lifeworld *replaces* the normal spatial determinants of identity.

This is a momentous change because, as Erikson (1970: 229f.) explains, identity formation in primitive cultures, as evidenced in the Yurok and Sioux tribes, was a process of the human subject growing within and as part of the surrounding environment. "Body and environment, childhood and culture may be full of dangers, but they are all one world. This world may be small, but it is culturally coherent." Similarly, Jung (1968: 102) explains that place forms part of the mother archetype; a cornerstone of the collective unconscious and "the form into which all experience is poured." He describes the mother archetype as:

> Many things arousing devotion or feelings of awe, as for instance the Church, university, city your country, heaven, earth, the woods, the sea or any still waters, matter even, the underworld and the moon, can be mother-symbols. The archetype is often associated with things and places standing for fertility and fruitfulness: the cornucopia, the ploughed field, a garden. It can be attached to a rock, a cave, a tree, a spring, a deep well, or to various vessels such as the baptismal font, or to vessel-shaped flowers like the rose all the lotus.
>
> (Jung 1968: 81)

Today, it appears that identity formation is decreasingly located in any space-time location, where *matter matters* and traversing space takes time, and increasingly in the stream of phantasmagorical experiences offered by cybernetics. This is especially true for individuals in Generation Z, who are growing up within, and as part of, the ever-expanding cybernetic realm. For example, it appears that some adolescents shape their identity by intentionally modifying their online personality and then integrating the feedback they receive into their identity (for a review, see Wood et al. 2016). Of course, "when we use devices to represent ourselves, we rely on what the devices are able to measure" (Rettberg 2014: 75). Therefore, the effect of cybernetics on identity formation in Generation Z may be twofold: disconnecting the individual from the real, where identity formation has historically taken place, and obstructing an accurate self-representation to others, leading to inaccurate feedback, which is in turn used in a faulty identity formation process.

Stage 3: technology to see with

Beyond obliterating the limitations of distance and space, cybernetics is now beginning to mediate the perception of the immediate physical environment. This is evident in emerging technologies that *see* on behalf of the human subject, including global positioning system devices (GPS), self-driving cars, personal drones, and augmented reality. Currently, augmented reality in particular is rapidly developing in areas as diverse as education (for a review, see Akçayir and Akçayir 2017) and surgery (for a review, see Kolodzey et al. 2016).

Virilio (1994: 61) broadly refers to these technologies as "vision machines." With vision machines, the individual is not only in the non-place of cybernetics, but is reduced to a *partial-self*, no longer even directly experiencing the immediate geographical space. Here, "images and representations replace the real, the object of representation declines in importance, and a domain of images and digital representations replace reality" (Kellner 2000: 110). Therefore, vision machines add to the mediated lived experience created by the cybernetic realm, further distancing the human subject from the real. This technology brings Baudrillard's dystopian vision of the future closer to life: "There is no separation any longer, no empty space, no absence: you enter the screen and the visual image unhindered. You enter your life and as you walk onto a screen. You slip on your life like a data suit" (Baudrillard 2002: 177).

Vision machines make further changes to the phenomenology of perception because here, the lived experience of the immediate physical environment is itself based on the pure light of the pixel. Therefore, perception reaches a zero point because pure light itself is devoid of meaning—there is nothing at all to be *seen*, even in front of you. The only way the individual can make sense of what is given to perception is by using pre-existing knowledge structures to interpret what is perceived. New experiences are not so much *created*, as old ones are *recycled*; the individual is stranded in a perceptually empty space, merely remembering class I reality from behind the miasma of machine-mediated perception. It is interesting to note, however, that Baudrillard (2005: 78) does not believe the pure light of the pixel even serves as a signifier any longer. He explains that people, "do not see the message; they only see the image … [people] believe only in the ascendency of signs. They long ago said goodbye to reality. They have gone over, body and soul, to the specular."

For Generation Z in particular, vision machines may significantly influence the identity formation process. By obscuring the immediate physical environment, vision machines appear to alter the very perception of the building blocks of identity normally received from direct interactions with the immediate physical environment. For example, an individual walking through a park using an augmented reality device may miss phenomena that he or she would have perceived if it were not for the device. Although phenomena such as feeling the vicissitudes of the breeze on one's skin or the texture of the ground under one's feet may be processed unconsciously, it is these very experiences that help locate the human subject as a being *in* the world and as an identity with fixed coordinates and dimensions. Cubitt (2011: 71f.) explains that this kind of loss of perception "throws into doubt the very capacity for being which humans derive from their experience of the world … not only do we no longer perceive the world, but it no longer perceives us."

Evidence of the dislocating effect of vision machines is appearing in news stories reporting bizarre incidents of individuals becoming lost in space and divorced from their bodies and the social world. For example, in San Diego, two individuals playing the augmented reality game, Pokémon Go, unwittingly walked over a 50ft cliff in their pursuit of Pokémon (Cockburn 2016). In another example, a man who was stabbed by an attacker while he played Pokémon Go chose to continue playing the game rather than seek medical attention in the hope that he could catch more Pokémon (Horton 2016). Therefore, in the same way as Baudrillard (1995: 61) claimed, "the Gulf War did not take place," because what was broadcast in the news was a simulacrum, it can be said that in the cybernetic realm of the vision machine, "reality itself does not take place," because what is given to consciousness is a simulacrum.

Similarly, a recent study reported that the use of GPS devices fundamentally, "alter[s] how people interpret, learn, navigate, and experience space and places" (Leshed et al. 2008: 6). The researchers found that GPS devices lead users to experience an abstract representation of the world, "reducing the need to feel oriented, keep track of locations, and maintain social interactions regarding navigational issues … inhibiting the process of experiencing the physical world by navigating through it" (ibid.). Without navigating *through the world*—without interacting with the real itself—the human subject is frictionless, phenomenologically *off-world*. Here Merleau-Ponty's observation appears to no longer hold true: "Our body is in the world as the heart is in the organism: it keeps the visible spectacle constantly alive, it breathes life into it and sustains it inwardly, and with it forms a system" (Merleau-Ponty 2002: 235).

We can see the importance of experiencing our body *in the world* in Jung's (1989) book, *Memories, dreams, reflections*. After his traumatic split with Freud, Jung (ibid.: 173) reflects:

I consciously submitted myself to the impulses of the unconscious. The first thing that came to the surface was a childhood memory from perhaps my tenth or eleventh year. At that time I had a spell of playing passionately with building blocks. I distinctly recalled how I had built little houses and castles, using bottles to form the sides of gates and vaults. Somewhat later I had used ordinary stones, with mud and mortar.... To my astonishment, this memory was accompanied by a good deal of emotion. "Aha," I had said to myself, "there is still life in these things."

Soon after this realization, Jung (1989: 174) began accumulating stones and started building cottages, a castle, and eventually a whole village. Every day after lunch, Jung continued his "building game" until his patients arrived. If he finished his work early, he also played in the evenings. The effect on Jung was profound. "In the course of this activity my thoughts clarified and I was able to grasp the fantasies whose presence in myself I dimly felt." For the rest of his life, Jung returned to this building game after traumatic events, including after his wife's death. Jung (ibid.: 175) explains, "The close of her life, the end, and what it made me realise, wrenched me violently out of myself. It cost me a great deal to gain my footing, and contact with stone helped me."

It is as if playing with stones allowed Jung to relocate himself as a subject *in* the world. This therapeutic effect appeared to come from his direct contact with the immediate physical environment, from returning to the real itself. It is therefore interesting to consider what effect proliferating vision machines will have on Generation Z. Indeed, if human psychology is dependent on contact with the world—if it is located within the coordinates of the real—what will happen as that contact is increasingly mediated by cybernetics?

Stage 4: technology to see *you* with—inner non-place

The riddle of technology ... is the riddle of the accident ... every technology produces, provokes, programs a specific accident. For example: when they invented the railroad, what did they invent? An object that allowed you to go fast, which allowed you to progress—a vision à la Jules Verne, positivism, evolutionism. But at the same time they invented the railway catastrophe.

(Virilio and Lotringer 2008: 45f.)

The most recent technological developments indicate that the cybernetic realm is beginning to infiltrate the human subject. Moore's law states that the number of transistors in a dense circuit approximately doubles every two years (Moore 1965). Today transistors crammed onto integrated circuits have become so small that they are capable of being inserted into the brain. For example, researchers using brain machine interface (BMI) have already shown that monkeys (*Macaca mulatta*) implanted with intracortical microelectrode arrays in the motor cortex were able to control a prosthetic arm with their thoughts (Velliste et al. 2008).

These primates quickly learned how to feed themselves and even lick the gripper portion of the prosthesis. Virilio referred to this cybernetic infiltration as the "endo-colonisation of the body," and explains that, "the human body is eaten up, invaded, and controlled by technology" (Armitage 2000a: 50). More recently, Žižek (2010: 334f., emphasis in original) reflected on how this invasion may end movement altogether:

> Even Stephen Hawking's proverbial little finger—the minimal link between his mind and the outside world, the only part of his paralysed body that he can move—will thus no longer be necessary: the mind will be able *directly* to cause objects to move, it is the brain itself which will serve as the remote control machine.

However, it appears that even beyond remote-controlled movement, phenomenological apperception of the self may end. "The prosthesis is *completely* alienating" (Virilio and Lotringer 2008: 88, emphasis added). Here, the immediate-organic self-experience is no longer machine-mediated but nullified altogether. Žižek (2010: 344f.) explains:

> The prospect for the near future is the explosive development of direct links between computers (and other media) themselves, which will then communicate, make decisions, etc., on our behalf, and simply present us with the final results of their interaction.

For example, as the size of the transistor is squeezed toward the infinitely small (that is less than a 50 micrometer diameter), self-communicating cybernetics such as "neural dust" are being built to literally run the body (Seo et al. 2016: 529). This technology is beyond Virilio's cybernetic class *n* realities. Here, there is no substitute category of reality. Instead, the individual has started to be *replaced*, exposing the danger of a "slide into a future without humanity" (Armitage 2000b: 13). To cease functioning at this zero-point of class I reality is to lose all dimensionality. Here, the pure light of the two-dimensional screen is long gone, as is the apperception of one's corporeal existence itself. "The prosthesis control[s] us in the very heart of our self-experience" (Žižek 2010: 341).

A large part of Virilio's work focuses on the invention of the accident, which he defines as the negative potentiality inherent in every technology (see Virilio 2007). The accident is therefore the realization—the *real discovery*—of the invention's true character. For Virilio, the cybernetic accident is the replacement of the physical dimension through the *complete* removal of the individual from class I reality. The cybernetic accident is absent existence. For Virilio, this is a traumatic historical event leading to the phenomenological "decomposition" of reality. Here we are in the midst of an ontological accident (Virilio and Lotringer 2002: 141). Therefore, whereas space was once the external referent of being in the world (Merleau-Ponty 2002), the infinitely small spaces within the human subject appear to have become the location of the loss of *being itself*.

We hypothesize that, for Generation Z, this kind of cybernetic accident will represent an almost total loss of identity; it is not only a loss of the geographical self as defined by Matless (1997), but also it is a loss of the existing *inner self* as a human subject. Indeed, we wonder whether Generation Z will be the first generation to experience the self as a non-place. Baudrillard demonstrated penetrating insight here in his reflection on the compact disc, one of the first frictionless products that did not need to interact with anything to function, including itself: "It doesn't wear out, even if you use it. Terrifying. It's as though you never used it. So it's as though you didn't exist. If things don't get old any more, then that's because it's you who are dead" (Baudrillard 1996: 32f.). For Generation Z, this metaphysical death may be more acute, for instead of not using *things* they may be the first generation not to use the self.

Conclusion: the return of the repressed

We are all dissidents of reality today, clandestine dissidents most of the time.

(Baudrillard 2005: 36)

The increasing speed of the modern technology—from cybernetics seeing the human subject's immediate environment for them to cybernetics moving inwardly, within the self—appears to be leading to the accident of a widespread loss of human identity, the disappearance of the geographical as well as inner self. Indeed, Merleau-Ponty's (2002) very hypothesis that the lived experience constitutes the interconnection of mind, body, *and* world, is changing. Virilio explains that, "Today we are no longer truly *seers* (*voyants*) of our world, but merely *reviewers* [*revoyants*]" (Virilio 2005: 37). We no longer see the world from within it, as a fundamental part of it, but at a distance. As a result, the world no longer sees us either, nor does it shape our identity. This is increasingly true for the Generation Z, for who the map may truly precede the territory (Baudrillard 1994).

We believe that evidence of this unfolding accident can be found in behaviors signaling the return of the repressed (Freud 2001a, 2001b). Freud hypothesized that the return of the repressed is the process whereby unacceptable unconscious memories try to force their way through to consciousness by reappearing as over-determined and often unrecognizable thoughts, symptoms, and behaviors, often leading to a profound internal struggle. "[The patient] is obliged to *repeat* the repressed material as a contemporary experience instead of, as the physician would prefer to see, *remembering* it as something belonging to the past" (Freud 2001a: 18, emphases in original). Examples of the return of the repressed include parapraxes, fantasies, and neuroses. Examples relevant to the loss of identity can be observed at the societal level in geopolitics and at the individual level in emerging psychopathologies.

At the societal level, we believe the return of the repressed is evident in the re-emergence of nationalist politics around the world (see *The Economist* 2016). In 2016, Americans elected Donald Trump president. Part of his campaign promised to build a wall at the Mexican border. Earlier that year, across the Atlantic, Britons voted to leave the European Union. Part of the "Brexit" campaign included posters depicting Middle Eastern immigrants clamoring to get into Britain. At the same time in Europe, popularity increased for nationalist, anti-immigration leaders, including, but not limited to, Marine Le Pen, leader of the Nationalist Front party in France and Geert Wilders, leader of Party for Freedom in the Netherlands. In response to these events, *The Economist* (2016: 51) remarked that, "It is troubling … how many countries are shifting from the universal, civic nationalism toward a blood-and-soil, ethnic sort … solidarity is mutating into distrust of minorities." We hypothesize that this phenomenon is not so much a simple, direct revolt against migrant crises, but at least in part, the return of the repressed loss of identity; an involuntary irruption of the repressed knowledge that human identity has *already* been lost to the cybernetic realm. Whether Generation Z experiences this loss of identity—whether they have the opportunity to even create such memories to repress—remains to be seen.

At the individual level, we believe the return of the repressed is evident in the increased prevalence of chronic and problematic shyness; this has been defined as, "A heightened state of individuation characterised by excessive egocentric preoccupation and overconcern with social evaluation, … with the consequence that the shy person inhibits, withdraws, avoids, and escapes" social situations (Zimbardo 1982: 467f.). Between 1977 and 2007, self-reported shyness increased from 40 percent to 58 percent (Carducci et al. 2007). We hypothesize that this phenomenon is not simply caused by the increased of negative emotional states and private self-consciousness (Henderson et al. 2014), but, at least in part, the compulsion to repeat the repressed loss of identity. Indeed, when the human subject has been disconnected from the real—when the body itself has been displaced as the center of the lived experience—it is possible that the unconscious repeats the repressed through a total disconnection with the other. Therefore, as proliferating class *n* realities replace class I reality, we believe young people in particular will increasingly experience chronic and problematic shyness, as the unconscious tries to force the repressed loss of identity through to consciousness.

In conclusion, in the cybernetic realm the individual (increasingly motionlessly) moves from one class *n* reality to the other, without interacting with the real, severing the connection with place *itself*—a great and vital component of the mother archetype (see Jung 1968). It can therefore be said that in the past, it was place that metamorphosized within the individual, co-constructing identity, but today, "perhaps you are indeed merely the machine's space now—the human being having become the virtual reality of the machine" (Baudrillard 2005: 80). If this is true, it is difficult to see how such a fundamental change will not profoundly influence human psychology, especially the still-developing identity of individuals in Generation Z.

References

Akçayir, M. and Akçayir, G. (2017) "Advantages and challenges associated with augmented reality for education: a systematic review of the literature," *Educational Research Review*, 20: 1–11. http://dx.doi.org/10.1016/j.edurev.2016.11.002.

Archer, K. (2013) *The city: the basics*, London: Routledge.

Armitage, J. (2000a) "From modernism to hypermodernism and beyond: an interview with Paul Virilio" in Armitage, J. (ed.) *Paul Virilio: from modernism to hypermodernism and beyond*, London: Sage, 25–55.

Armitage, J. (2000b) "Paul Virilio: an introduction" in Armitage, J. (ed.) *Paul Virilio: from modernism to hypermodernism and beyond*, London: Sage, 1–23.

Augé, M. (1995) *Non-places: an introduction to supermodernity*, London: Verso.

Baudrillard, J. (1994) *Simulacra and simulation*, Ann Arbor: University of Michigan Press.

Baudrillard, J. (1995) *The Gulf War did not take place*, Sydney: Power Institute.

Baudrillard, J. (1996) *Cool memories II: 1987–1990*, Cambridge: Polity Press.

Baudrillard, J. (2002) *Screened out*, London: Verso.

Baudrillard, J. (2005) *The intelligence of evil: or the lucidity pact*, Oxford: Berg.

Baudrillard, J. (2006) *Cool memories V: 2001–2004*, Cambridge: Polity Press.

Bruegel, P. (1559) *Netherlandish proverbs*, Berlin: Gemäldegalerie.

Carducci, B. J., Stubbins, Q. A. and Bryant, M. R. (2007) "Still shy after all these (30) years," *Paper presented at the American Psychological Association 115th National Conference*, Boston, MA.

Cockburn, H. (2016) "Pokémon Go: two men fall more than 50 ft off cliff while playing mobile game," available online at: www.independent.co.uk/news/world/americas/pokemon-go-men-fall-off-cliff-san-diego-android-ios-app-a7136986.html (accessed July 10, 2017).

Cubitt, S. (2011) "Vector politics and the aesthetics of disappearance" in Armitage, J. (ed.) *Virilio now: current perspectives on Virilio studies*, Cambridge: Polity Press, 69–91.

Debord, G. (2008) *The society of the spectacle*, New York: Zone Books.

Economist, The (2016) "League of nationalists," available online at: www.economist.com/news/international/21710276-all-around-world-nationalists-are-gaining-ground-why-league-nationalists (accessed July 10, 2017).

Erikson, E. (1970) *Childhood and society*, Harmondsworth: Penguin.

Freud, S. (2001a) *Beyond the pleasure principle, group psychology, and other works, the standard edition of the complete psychological works of Sigmund Freud Vol. XVIII*, London: Vintage.

Freud, S. (2001b) *Moses and monotheism, an outline of psycho-analysis, and other works, the standard edition of the complete psychological works of Sigmund Freud Vol. XXIII*, London: Vintage.

Guy, F. (1819–1820) *Winter scene in Brooklyn*, New York: Brooklyn Museum.

Henderson, L., Gilbert, P., and Zimbardo, P. (2014) "Shyness, social anxiety, and social phobia" in Hofmann, S. G. and DiBartolo, P. M. (eds) *Social anxiety: clinical, developmental, and social perspectives*, London: Academic Press, 95–115.

Horton, H. (2016) "Pokémon Go addict stabbed while playing, refuses to get treatment so he can continue," available online at: www.telegraph.co.uk/technology/2016/07/13/pokmon-go-addict-stabbed-while-playing-refuses-to-get-treatment/ (accessed July 22, 2017).

Husserl, E. (1970) *The crisis of European sciences and transcendental phenomenology: an introduction to phenomenological philosophy*, Evanston: Northwestern University Press.

Ihde, D. (1974) "The experience of technology: human-machine relations," *Cultural Hermeneutics*, 2: 267–279.

Jung, C. G. (1968) *The archetypes and the collective unconscious, the collected works of C. G. Jung Vol. 9 (Part 1)*, London: Routledge & Kegan Paul.

Jung, C. G. (1989) *Memories, dreams, reflections*, New York: Vintage.

Kellner, D. (2000) "Virilio, war and technology" in Armitage, J. (ed.) *Paul Virilio: from modernism to hypermodernism and beyond*, London: Sage, 101–126.

Kolodzey, L., Grantcharov, P. D., Rivas, H., Schijven, M. P., and Grantcharov, T. P. (2016) "Wearable technology in the operating room: a systematic review," *British Medical Journal Innovations*, 3: 55–63. http://dx.doi.org/10.1136/bmjinnov-2016-000133

Leshed, G., Velden, T., Rieger, O., Kot, B., and Sengers, P. (2008) "In-car GPS navigation: engagement with and disengagement from the environment," available online at: www.cs.cornell.edu/~tvelden/pubs/2008-chi.pdf (accessed: July 22, 2017).

Liebenberg, L. (2006) "Persistence hunting in modern hunter-gatherers," *Current Anthropology*, 47 (6): 1017–1026.

Matless, D. (1997) "The geographical self, the nature of the social, and geoaesthetics: work in social and cultural geography, 1996," *Progress in Human Geography*, 21 (3): 393–405.

Matless, D. (2016) *Landscape and Englishness*, London: Reaktion Books.

Merleau-Ponty, M. (2002) *Phenomenology of perception*, London: Routledge.

Moore, G. E. (1965) "Cramming more components onto integrated circuits," *Electronics*, 38 (8): 114–117.

Rettberg, J. W. (2014) *Seeing ourselves through technology: how we use selfies, blogs, and wearable devices to see and shape ourselves*, New York: Palgrave Macmillan.

Seemiller, C. and Grace, M. (2016) *Generation Z goes to college*, San Francisco: Wiley.

Seo, D., Neely, R. M., Shen, K., Singhal, U., Alon, E., Rabaey, J. M., Carmena, J. M., and Maharbiz, M. M. (2016) "Wireless recording in the peripheral nervous system with ultrasonic neural dust," *Neuron*, 91 (3): 529–539.

Shea, J. J. (2006) "The origins of lithic projectile point technology: evidence from Africa, the Levant, and Europe," *Journal of Archeological Science*, 33 (6): 823–846.

Thieme, H. (2000) "Lower paleolithic hunting weapons from Schöningen, Germany: the oldest spears in the world," *Acta Anthropologica Sinica*, 19: 136–143.

Urry, J. (2007) *Mobilities*, Cambridge: Polity Press.

Velliste, M., Perel, S., Spaling, M. C., Whitford, A. S., and Schwartz, A. B. (2008) "Cortical control of a prosthetic arm for self-feeding," *Nature*, 453 (7198): 1098–1101.

Virilio, P. (1991) *Lost dimension*, New York: Semiotext(e).

Virilio, P. (1994) *The vision machine*, London: British Film Institute.

Virilio, P. (2000a) *A landscape of events*, Princeton: Princeton Architectural Press.

Virilio, P. (2000b) *The information bomb*, London: Verso.

Virilio, P. (2005) *Negative horizon: an essay in dromoscopy*, London: Continuum.

Virilio, P. (2007) *The original accident*, Malden: Polity Press.

Virilio, P. (2009) *The aesthetics of disappearance*, Los Angeles: Semiotext(e).

Virilio, P. and Lotringer, S. (2002) *Crepuscular dawn*, New York: Semiotext(e).

Virilio, P. and Lotringer, S. (2008) *Pure war*, New York: Semiotext(e).

Wood, M. A., Bukowski, W. M., and Lis, E. (2016) "The digital self: how social media serves as a setting that shapes youth's emotional experiences," *Adolescent Research Review*, 1 (2): 163–173.

Zimbardo, P. G. (1982) "Shyness and the stresses of the human connection" in Goldberger, L. and Breznitz, S. (eds) *Handbook of stress: theoretical and clinical aspects*, New York: Free Press, 466–481.

Žižek, S. (2010) *Living in the end times*, London: Verso.

7 From map reading to geobrowsing

Methodological reconsiderations for geomedia

Pablo Abend

Prologue: browsing through magnetic cows

At the end of the 1980s, Hynek Burda made a spectacular observation. The research fellow at the institute for zoology at the Johann Wolfgang Goethe University in Frankfurt had been concerned with the ethology of subterranean rodents (Fukomys anselli) when he noticed symmetric nest-building patterns of animal families inhabiting isolated enclosures in the laboratory. After some thinking and more observations, Burda concluded that the placement in the Southeast of the enclosure must follow an overriding principle, which led up to the working hypotheses that the earth's magnetic field accounts for this behavior. While a magnet sense of migrating birds, pigeons, tortoises, and common whitethroats among others had been repeatedly proven in zoological laboratories ever since the 1960s, no one had ever done so with small mammals thus; Burda wanted to close this research gap. Therefore, he conducted an ethological experiment in a shed at the botanical gardens in Frankfurt on the Main. Inside the shed he built a round arena, covered the ground with soil, and laid out pulp as building material for the rodents. A Helmholtz coil above this nesting place made for an artificial magnetic field, the polarity of which could be changed in direction by rotating the device. Once everything was ready, different families of the eusocial creatures were deployed in the arena and started to build their nests in the Southeast of the artificial habitat, just like they did in the laboratory. The subsequent adjustment of the Helmholtz coil confirmed the hypotheses; when the coil was rotated and magnetic North changed, the rodents abandoned their old nests in order to build new ones in the artificially produced Southeast. Burda provided sufficient proof for the magnetic alignment of subterranean rodents (see Burda et al. 1990). Years later, Sabine Begall joined Burda, now Chair of General Zoology at the University of Duisburg-Essen, in his research. They continued to work with the rodents and Begall managed to locate the organ sensitive to magnetoreception in the vestigial eyes of the nearly blind creatures. However, they had been thinking all along about studying the alignment of larger mammals such as cows and horses but they were not able to figure out the means to do so. Certainly, a Helmholtz coil this big could not be built, an observation in situ would be too costly, and satellite imagery was inapt since the cattle had to be located first.

Then, during a PhD exam in 2008, the zoologists came up with an idea: Why not use the freely available digital globe software, Google Earth, to search for grazing cattle all over the world? They immediately tested the method, and after finding 8150 cows on 308 randomly selected pastures on the satellite images, Begall and her team empirically validated and statistically verified a magnetic alignment of cattle toward geographic North. Nature.com published the initial story and a widely recognized article in the *Proceedings of the National Academy of Sciences* (Begall et al. 2008) and a second paper that served as proof for the first (Burda et al. 2009) followed. The publication sparked an ongoing controversy that started in the comment section underneath the blog post and continued in university departments and laboratories. The majority of the critical voices were aimed at the validity of the use of Google Earth in this type of research. The suitability of the geobrowser as a substitute for research trips or for laboratory work is called into question. Commentators argue that Google Earth has not been specifically produced for the purpose of scientific processing and only shows an idealized segment of reality and others state that this kind of desktop zoology cannot replace the laboratory; whereas, for some commentators, Google Earth proves to be more valid because of the possibility to extinct disruptive factors.

Besides the inconsistencies within the argumentation of the critics, this episode explicates several issues that research into the nature and usage of digital maps is facing today. It touches on topics such as the objectivity, functionality, and usage of digital geomedia while questioning the representational status of the object. Is it still safe to speak of a specific cartographic indexicality or do we need a wider framework for the study of maps? Is indexicality a crumbling commodity in the market for geographic information? The first part of the chapter addresses these issues by using the concept of the map as an "immutable mobile" (Latour 1986, 1987) as a starting point. I then show how new actors enter the market for geographic information and how mutable elements are introduced to maps. This part of the chapter follows the hypothesis that the digitalization of cartography has resulted in a radical shift that has fundamentally altered "the cartographic gaze" (see Pickles 2004: 80ff.) of the modern era through changes in the relation between map user and cartographic presentation. It is argued that this shift is brought forward by practices of "geobrowsing" (Peuquet and Kraak 2002; Abend 2013) that draw into doubt whether map use can still be modeled as a reading process. Consequently, the last part of the chapter focuses on some methodological challenges for the study of map use and for the role of geomedia in shaping our understanding of space and place.

Maps as immutable mobiles

For Bruno Latour (1986, 1987), maps are prototypical "immutable mobiles," or more precisely "immutable and combinable mobiles" (Latour 1987: 227). He introduces the heuristic concept in order to explain the vast accumulation of knowledge in the modern age. Latour does not limit the concept to cartographic media, but rather, he repeatedly takes the map as an example and instance of map production as illustrations for the accumulation of scientific facts by means of creating immutable mobiles. Originally, Latour used the immutable mobiles to describe the evolution of modern-day science whereby he focused on techniques such as writing and visualizing, which produce objects such as, "paper, signs, printings, and diagrams" (Latour 1986: 3). This refers to all kinds of media, also including transportation technologies, which were invented to collect facts on one place and shift them to some other site where they can be assessed and combined. Thus, cultural techniques such as geometrization and proportional projection lay at the heart of scientific work that transforms matter into form (see Latour 1999: 69) and has to secure a "consistency of form across transformations" (Schüttpelz 2009: 72, translation by the author). This transformation of knowledge follows a strategy of deflation through which knowledge acquired by means of observations, measurements, and empirical methods is transformed into two-dimensional inscriptions that can be combined and superimposed with subsequently acquired facts and transported through various social contexts. The mobilization of facts is core in this model (Latour 1986: 7) and the cartographic visualization is a prototypical result of the deflation, combination, superimposition, and mobilization of facts about the world and their rendering in a presentable form.

In his seminal work *Science in Action* (1987), Latour uses the example of cartography for the first time to develop his idea of the immutable mobile by means of a historical anecdote: In 1787, the French King Louis XVI sent out Jean-François de Galaup de Lapérouse on an expedition to Sakhalin in China in order to bring back a map. This map was supposed to resolve the controversy about whether Sakhalin is an island or a peninsular. After Lapérouse's arrival in Sakhalin, he asked an old native to draw a map in the sand at the beach. This ephemeral drawing, threatened by the water, was conveyed into the notebook of the explorer so that it could be returned to France and shown to the king who was then enabled to act upon it and use this knowledge for subsequent explorations. The moment of transferring the map drawn in the sand onto a piece of paper is crucial for Latour because this act of rendering the facts durable secures their safe passage of the morphology of Sakhalin to Europe. Only the shifting of the carrier medium made it possible to make the local geographic knowledge of the inhabitants in the Pacific accessible to the king and his geographers in far away France. As a consequence, this not only resolved a scientific dispute about the morphology of the far away territory but also the local knowledge could be used as a basis for further explorations and possibly exploitations in the area. The production of immutable mobiles lays at the heart of the initiation and maintenance of the circle of knowledge and eventually capital accumulation through which the knowledge about things in the periphery could be brought back to "centers of calculation" (Latour 1987: 215ff.), in this case congruent with the centers of colonialism. This knowledge can be used to exert power over the remote territory which is why immutable mobiles as networks of people and things also secure the means of "long distance control" (Law 1986) for central macro actors. Therefore, immutable mobiles are found particularly in situations in which actors are tangled up in an antagonistic situation they want to resolve in their favor.

Immutable mobiles are inscriptions that transform and translate the world in order to circulate knowledge. However, as the above-mentioned episode with the zoologists using Google Earth illustrates, immutability is not a given but it has to be carefully produced and knowledge about magnetic cows has to pass through a whole chain of interlinked inscriptions (see Latour 1999: 24ff.) until it becomes a published scientific fact. The zoologists had to perform a number of tasks and bring about a range of transformations in order to generate publishable zoological facts. In the heat of the controversy that followed the publication, this chain of operation hidden in the journal articles as end products of their knowledge production had to be disclosed to the public. In a document called "Lesson for sceptics," Burda and his colleagues offer sample data, show how they have been extracting images out of Google Earth via screenshots, and describe how these images are combined with different instruments such as a circular scale and an on-screen distance meter. All along, reference is secured through a chain of interlinked inscriptions through which the knowledge of the cows passes. The zoologists show how they produce their scientific facts by using people, signs, and objects and hereby managed to calm the critics. They reverse the sequence (see Latour 1999: 61) and the site of dispute shifts from the public blog to scientific journals with several teams trying to reproduce the research results.

What the zoologists cannot disclose however, are the technological prescriptions (Akrich and Latour 1992) of Google Earth in the first place which work as a precondition of their work. The interface of the geobrowser is not simply an object that can be traced back to local facts, at least not to the facts the zoologists are interested in (the state of the cows) and there is reasonable doubt that such an ideal immutable mobile built on the gift of local knowledge is as easily conceivable as the Lapérouse episode makes one think (see Bravo 1999). Instead, the artifact the zoologists have been attaching their work onto is in itself made up of a long chain of intermediaries and mediators from the satellites collecting sensor data via the algorithms producing images and stitching them together to the presentation of the navigable globe on the screen of the end user.

Introducing mutability to maps

Several authors point out that the generally assumed immutability of maps is challenged by the digital transformation and the rise of geomedia. Scholars of geography and GIS experts alike have been referring to distinct changes in the modes of cartographic production (e.g., Caquard 2013, 2014; Gartner 2009; McHaffie 1995; Pickles 1995). Meanwhile, commercial and non-profit mapping applications have become aesthetically enticing by opening platforms which transform previously abstract, logocentric, and disembodied cartographic maps into egocentric and embodied action spaces by combining or overlaying with photographic and sensorial images and 3D computer graphics. In addition, a gradual move from capital-intensive authoritarian modes of production lowers the bar for cartographic amateurs to get involved in map design and diffusion. Participatory and collaborative forms of cartographic cultures emerge online (Caquard 2014; Goodchild 2007; Turner 2006), and as a result, digital maps with personal content as well as local and micro-local knowledge have become available privately and publicly. Two major shifts in the cartographic mode are of interest here: the appearance of new actors on the market for cartographic products and the accompanying transformations of the artifacts referred to as geomedia which challenge the notion of the map as an immutable mobile.

Bearers and administrators of geographic knowledge are no longer governmental institutions and research facilities. Thus, professional geographers and cartographers are confronted with actor-networks outside their subject areas. Increasingly, producers of mobile devices, social network providers, and open source communities collect, assess, and distribute geographic data. Non-profit organizations such as the OpenStreetMap integrate crowdsourced local and hyperlocal information with the help of GPS-enabled neogeographers. Commercial services like Google Maps integrate user input through easy-to-use services that afford geotagging data, annotating locations, and publishing geographic information. In recourse on the circles of accumulations one could sum up that geographic facts are no longer stored in particularly easy to spot centers of calculations (e.g., the archives and data banks in Paris as a center of European Enlightenment or the research facilities and federal land registry offices of the modern age) but are distributed in networked data centers all over the globe. As a result, accumulation circles multiply with small-scale loops that feed back into distributed data centers. This reproduces the innovation paradigm of the post-Fordist digital economy where objects are designed to allow permanent interactions with the consumers to secure (capital) accumulation in the network economy through a now dominant mode of "productive communication" (Hardt 1999: 94). Thereby, like other software as a service, most geomedia remain in a kind of perpetual beta with users constantly improving and complementing the database through their explicit or implicit participation (see Schäfer 2011: 77ff.) and heterogeneous data is made compatible in order to transfer it from one service to the next. For example, Google's augmented reality game *Ingress* (developed by Niantic, released in 2013) is based on Google Maps data and uses the original service's base map but it also collects additional user data which feeds back into this data base and is then used in the subsequent location-based game *Pokémon Go* (also developed by Niantic, released in 2016). In addition, mutable elements are added to otherwise immutable structures to form commodities that afford the adaptability of the same object to different contexts and usage scenarios (see Thrift 2006: 287). This transforms maps into what Star and Griesemer (1989) describe with the term "boundary objects." These are objects "which are both plastic enough to adapt to local needs and the constraints of the several parties employing them, yet robust enough to maintain a common identity across sites" (ibid.: 393).

Likewise, the design and functionality of many geomedia applications can be specifically adapted for different tasks across various and diverse social fields according to situated requirements. For example, Google Inc. widely promoted Google Earth as a means of virtual travel, but many other usage scenarios have been emerging long after the release date of the software. Some can be found in quite unexpected contexts as the example of the zoologists illustrates. Like Google Earth, most digital geomedia are "weakly structured in common use, and become strongly structured in individual-site use" (Star and Griesemer 1989: 393). As maps turn into services and become integrated in digital platforms they still rest with relative stability on indexical cartographic data (the base map or the grid) but can adapt through the principle of layering and annotation: they become more mutable than analog maps. In order to establish and secure data-driven innovation circles, the users themselves become part of the circles of data accumulation. However, a great share of the accumulated data still has to become rendered immutable, at least temporarily, in order to feed the accumulated local data back into global systems.

Cartographic egocentricity

In order to foster such an "interpretative flexibility" (Pinch and Bijker 1987: 40) of geographic media, there had to be certain changes in the way geographic information is visualized, arranged, and ordered. Perhaps the biggest shift introduced to the digital upheaval of cartographic production is the dissolution of logocentric visualizations into egocentric perspectives introduced by the early car navigation systems (see Thielmann 2007) and further popularized by mobile location-based services. This egocentric mode was introduced to mobile phone users toward the end of the 1990s. An early project was the so-called Digital City Project started at AOL. The Digital City was conceived as a user-centered location-based service that allowed content to be dynamically organized in a pre-determined radius around the user using a cartographic interface with the aim to show users the quickest route to products and services. At the time, it was not possible to turn this kind of user-centered service into a marketable product because the telecommunications infrastructure was not sufficiently developed to support the necessary data transfer. Nevertheless, the concept flowed into the development of services like Google Local that eventually became Google Maps.

With the popularization of smartphones, the idea would reach its full potential leading to a "mobile egocentricism" (see Thielmann 2007). In this egocentric cartographic mode, the viewer always remains in the center of the data space that dynamically orients and reorients itself around this center in every moment of movement. While the logocentric presentation is coercing the viewer apart from the frame of reference of an image externalizing the subject from the objects, the egocentricity of current mapping applications posits the subject (e.g., by means of a representation) inside the image where it shares the same space with the objects (see Klatzky 1998: 11). In the resulting geospatial data spaces, information is delivered to the position at which the user is waiting for it. Instead of a state of "being there" as the transport mode of media—commonly referred to as the feeling of (spatial, social, or temporal) presence (Lombard and Ditton 1997), involvement (Calleja 2011) or in terms of a technically induced immersion (e.g., Slater and Wilbur 1997)—the mode of "you are here" is established through permanent tracing of the subject and the provision of location-based data and information all at once. This leads to the circumstance that the conventional logocentric and mostly also ethnocentric orientation of maps is openly sacrificed to a flexible egocentrism that continually places subjects at the center of the (media) world. In this way the reference frame for map usage becomes more flexible, which has an overall effect on the organization of the data space itself. The geospatial web or geoweb (see Scharl 2007) is above all a subject-centered network that organizes information relevant for the localized user and reorganizes this information dynamically as soon as motion occurs. This shift from an absolute, logocentric to a relational and dynamic egocentric world order has not only led to new usage scenarios for cartographic media but also has perhaps even led to a whole new dispositive. For, as Eric Gordon (2009: 409) explains, in the light of geographical media, a paradigmatic revolution in the overall conception of the internet can be observed, which "transform one's experience of the network from something distant and external to something intimate and internal."

Whereas the concept of the immutable mobile fosters the cartographic mode of the logocentric, distant, and top-down view on a given territory, also referred to as the "god trick of seeing everything from nowhere" (Haraway 1988: 581) or as the illusion of the "cartographic gaze" (Pickles 2004: 80), the map as a boundary object gives way to egocentric, alternating, and angled perspectives. Thus, geomedia transforms maps into mutable artifacts in the sense that, through convergence with other media, a gestalt-switch according to the context and purpose of use is afforded. As a result, the transformation in the perception of the internet—once labeled by the apologist of the new era Tim O'Reilly as Web 2.0—is not primarily to be understood as a technological development but as a shift from a logocentric to an egocentric network spurred to a great extent by innovation and product strategies of companies on the market. Consequently, the egocentrism of current mapping applications bears witness to an overall reframing of the internet's "metageography" (Gordon 2009) in the age of the social web. Instead of remaining outside the territory, one is right in the middle of the media space. Additionally, with this position in the middle of the cartographic presentation, which is now interwoven with the network and cannot be separated from it, our body itself becomes a point of interest in the socio-technical grid. Geomedia are thus by nature localizing technologies.

Critiques have stated that this tendency toward egocentrism can be read as a symptom of a reorganization of space that follows the inscription of commercial interests within software by the producers (Zook and Graham 2007). According to Eli Pariser (2011), the platform providers tend to create spaces he has termed "filter bubbles." Filter bubbles can be described as topologically organized data spaces presenting users with an already short list that is condensed and curated using data-mining techniques. The content within the bubble is generated from the data shadow one leaves behind and consists of information flagged as related and similar to past search preferences. The egocentricity that prescribes the experience for the end user is used for commercial interests here and geomedia becomes part of much broader efforts to control and steer the "dwelling in the web" (Thielmann et al. 2012).

Geobrowsing: from a mimetic to a navigational understanding of the map

Traditional map practices associated with the cartographic gaze are not going away but, rather, they have been transformed and are now accompanied by different cartographic modes that cannot be described in terms of map reading alone. In addition, applications such as location-based games use ludocartographic interfaces that afford playful practices based on geographic data to further extend the notion of maps as tools. Additionally, when inscriptions are combined with 3D models and manipulation devices, they can transform into simulated geographies that are comparable to non-indexical media that allow the exploration of a vast terrain as open world computer games do. These practices can be subsumed under the term "geobrowsing" (Peuquet and Kraak 2002; Abend 2013). The notion of geobrowsing highlights geomedia's potential for ludic interactions with geographic space that break with established conventions such as the Northern orientation and the top-down view. Thereby, the term points to an explorative mode of mapping which complements and sometimes replaces spatial control by ludic interventions. Geobrowsing as a media practice falls somewhere between goal-oriented processes of information retrieval and a less targeted action related to consumerist media practices.

Kingsbury and Jones III (2009) draw on Nietzsche's distinction between the Apollonian and the Dionysian to seize on the dichotomous division of the map into an analytical and purposeful tool contrasted to a concept of the map as an affective medium, emphasizing the playful side of map use that characterizes the practice of geobrowsing. While critics stress the Apollonian understanding of the medium pointing out either anxiety or hope, the Dionysian, which manifests itself in the appropriation of the common users, offers an alternative interpretation, a mode of use that follows a flâneur-like and affective form of interaction while emphasizing "the indeterminacy of technology" (ibid.: 503).

Thus, the Apollonian gaze is closely tied to a power/knowledge complex Michel Foucault identified throughout his work: hegemonic discourse. Mapping as a political act is a duty given to the authorities, and they use maps within the exercise of control over national space, or, in the case of Digital Earth, commercial vendors inscribe their view in the alleged objective presentation. By contrast, Kingsbury and Jones III argue in recourse to Walter Benjamin's fascination with the figure of the flâneur that Dionysian forms of usage are supported by the new means of navigating within the interface of contemporary mapping applications. The Dionysian suggests the emergence of a space that fosters collaborative production and individual experiences of space and place. Using the example of Google Earth, the authors identify spontaneous, chaotic, and even avant-garde forms of spatial disclosure. They keenly focus on the discussions and image collections in blogs and forums where users share knowledge on technical errors, defocused or blurred images, and extraordinary discoveries as non-intended by-products of geographic visualization. While many of the examples can seem trivial, of interest in considering geobrowsers is the explicit inclusion of different forms of user interactions, along with the renunciation of strict ontological assumptions of the geomedial form in question. Technical errors such as bugs, glitches, or system failures by no means make every information transmission impossible, but they cause moments of emergence and openness that can be productive within the process of appropriation. While it is doubtful if a dichotomy is helpful for the research into the actual usage of geomedia, the examples show that online and interactive digital geomedia are no longer fully and satisfactorily describable as self-contained indexical sign systems, but take on a semantic openness arising in their under-determinacy that has to be taken into account (Abend and Harvey 2017). Actions are enabled by design and interface, but there is no longer an implicit user that can be fully modeled in advance. Instead, the style of geobrowsing can be modified in every virtual step of the flâneur.

The Dionysian is closely bound to the egocentric cartographic mode. Both are based on a navigational understanding of mapping, according to which geographic image production and the traversing of space are always conceived together (see Verhoeff 2012). The navigational is further described by Valérie November, Eduardo Camacho-Hübner, and Bruno Latour (2010). With regard to cartographic platforms, the authors argue that the reference of a map and thus its representational function is not a given fact but subject to a negotiation between producer, user, map, and territory. This negotiation has been at the heart of the process all along, but until digitalization, it has been hidden behind a mimetic understanding of the map brought forward by the historical paradigm of objectivism. Only with digital cartography does the process of generating a reference between map and territory become visible (November et al. 2010: 286). The authors state that digital technologies alter our perception of cartographic images to the effect that a mimetic interpretation recedes in favor of a navigational understanding (ibid.: 582).

This distinction is further discussed in comparing analog and digital maps. While paper maps are looked at from above, a practice described as "looking at geographical data" (ibid.: 583), digital maps mean that the user has to log into a databank and actively engage with the presented material. Looking at the shift from static and discrete maps with a fixed single scale to dynamically adjustable, geographic platforms, the authors develop their argument by stating that the epistemological gap between map and territory can only be overcome step-by-step in interaction with the digital material. Mapping does not comprise of a leap from territory to map, or from objective to the semiotic world (see Korzybski 2005: 20, 269), but rather it shows a "deambulation" (November et al. 2010: 586) along leading signposts of a chain of inscription. Thus, the navigational bears the premise that inscriptions are not matched with some outer reality but only with previous and subsequent inscriptions. This forms a trajectory that one navigates along from world to map and vice versa. The value of individual sign-posts derives from previous experiences, but it is also bound to expectation structures. Thus, the quality of "guidance" becomes more important than the optical consistency of inscriptions that has been the primary characteristic of maps as immutable mobiles (see Latour 1986: 7f.).

Geobrowsing and the navigational emphasize the active role of the artifacts tangled up in the interplay of map, territory, perception, and experience. Mediated by cartographic egocentrism, the relational concept of the navigational allows for alternative viewpoints and perspectives of the world that do not seem as strongly preselected as in the times of fully immutable mobiles. The digital medium is supposed to leave enough clearance for the emergence of knowledge that has not been intended by the producers and not been expected by the users (see Peuquet and Kraak 2002: 87). It is this contingency that is supposed to turn geobrowsing into an instrument of reflexive learning and understanding. Seen this way, geobrowsing can be praised for fostering a navigational interpretation that turns the translation from map to territory into a tangible experience. In this view, geobrowsing becomes a practice that enables the user to follow through different layers and signposts, tracking changes and the dissemination of networks on the way. But such a relational conception of space also naturalizes the propositions of geographic space and masks the standards and protocols that prescribe the production and thus the experience of space and place. It reduces the distance to the objects and at the same time allows for the seamless movement through chains of inscriptions increasingly denying any differences stemming from the materiality of the inscriptions. This is what Latour (2013: 128, 323) has called "double-click information," which suggests that knowledge is immediately at hand while the transformation processes are invisible. In double-click mode, all protocols of data collection and accumulation remain hidden (Hind and Lammes 2016) and the chain of inscriptions is deleted with only the two ends of map and territory remaining. Through geobrowsing—when done with care— some technological conditions of reference production are rendered visible. At the same time, the egocentric mode of the presentation counteracts this aspiration in giving way to a flexible normalization of space that varies according to the location of use.

Some methodological reconsiderations for geomedia

It seems that this situation of the contemporary (geo)medial assemblage sharpens a situation that was already identified in the 1980s. Back then, Barbara Petchenik criticized the reductionism of the theoretical models of map reading that are based on system theory, the theory of communications, and cybernetics. She complained about "a lack of serious attention to the effect of motivation in naturalistic map use and to the nature of spontaneous task performance" (Petchenik 1983: 48f.) and pointed out that the Shannon-Weaver model of communication—even if it is enhanced by notions of system environment and cybernetic feedback loops—fails to grasp higher-level meaning because it reduces the user to a functional unit and the map to a system of interpretable signs. While the mainstream of cartographic education still holds on to the classical communication model (and also many critical cartographers still follow a map-as-text approach), this methodological claim is renewed in the face of ubiquitous cartographic media and their often complex and multilayered affordances. Perkins (2008: 152) updates the critique by focusing on the reductionism of communication models, which are in his view only insufficiently apprehending the context of map usage, and asks for more cultural research using qualitative methods such as ethnographic fieldwork, interviews, focus groups, and read-aloud protocols. In a similar fashion, Cartwright and Hunter (2001) see research confronted with user interface issues and draw attention to the possibility of emergent knowledge through interactions, Goodchild (2008: 20) declares with regard to Google Inc.'s globe software: "There is a need for fresh thinking ... that exploits the rich set of options available in Google Earth," while Dodge et al. (2009: 231) make the plea to leave the laboratory in order to grasp more "natural" forms of map usage.

It is rarely stated clearly that this implies rejecting traditional models of communication as well as questioning carto-semiotic approaches. However, if research into maps wants to keep track of the richness and complexity of geomedia, alternative frameworks to semiotic approaches and communication models have to be found. With geomedia creating their own medial spatial environments that invite us to position, contemplate, and play with geographic data, map use can no longer be classified as "map reading" but instead must be discussed as different modes of map viewing, navigating, charting, and indexing. The notion of geobrowsing hints to a shift from map reading to map usage in this broader sense. Empirically speaking, geobrowsing demands research efforts that turn away from semiotic approaches to focus on media practice instead in order to pose the question which geospatial configurations unfold efficacy for our everyday lives and how these configurations came into being (see Werlen 2008: 366). Geobrowsing happens within such configurations and at the same time brings new configurations into being. It happens within a "field of practice" (Schatzki 2006: 11) that is not only generated by discursive practices and communicative action but perhaps predominantly by concrete interactions with signs and material media technologies.

To focus on practices—led by the question what people do with media (see Couldry 2010: 38)—opens up the opportunity to look more closely at the cultural embeddedness of geomedia and reveal alternative attitudes toward mapping. This anthropological view, which takes user-sided appropriations of technologies into account, has to be accompanied by an artifact-centered analysis that addresses the material agency and its share in the spatial production process. Within such an actor-network, geospatial media can no longer be seen as passive intermediaries, but have to be addressed as active mediators, as a "mangle of practice" (Pickering 1995), that provides and limits a plethora of possibilities to choose, organize, and navigate geographic data. For the study of map usage, Cartwright and Hunter (2001: 302) only implicitly point to this when they mention the "intrinsic merit of the content," but there is also a need to look at other affordances of the media involved. Geomedia not only precede their territory (Pickles 2004) in being symbolic-ideological texts but they also socio-technically organize and reorganize our perception and conception of place and space (see Döring and Thielmann 2009). Thereby, focusing on concrete material practices highlights how mapping is intermingled with other modes of human experience which is in line with non-representational accounts of media use that highlight the procedural nature of reference generation and space production. In addition, looking at practices rather than the maps themselves helps to deal with a false exceptionalism of cartographic inscriptions that the immutable mobile concept might suggest by focusing the situated nature of spatial knowledge production rather than the properties of any map as a carto-semiotic sign system. The line between maps and everyday life is in no way manifested but is constantly negotiated between different ways of mapping out the territory. This requires scrutinizing the role of production and distribution technologies anew, whereby one would move—in line with the preceding excursus on the egocentricisms of the cartographic view—from the politics of the map text to the politics of placement and misplacement of subjects and data. It is essential that knowledge production is examined all the way from the production environment to the point where technology, text, and usage meet.

References

Abend, P. (2013) *Geobrowsing: Google Earth und Co – Nutzungspraktiken einer digitalen Erde*, Bielefeld: transcript.

Abend, P. and Harvey, F. (2017) "Maps as geomedial action spaces: considering the shift from logocentric to egocentric engagements," *GeoJournal*, 82 (1): 171–183. http://dx.doi.org/10.1007%2Fs10708-015-9673-z.

Akrich, M. and Latour, B. (1992) "A summary of a convenient vocabulary for the semiotics of human and nonhuman assemblies" in Bijker, W. E. and Law, J. (eds) *Shaping technology, building society: studies in sociotechnical change*, Cambridge, MA: MIT Press, 259–264.

Begall, S., Červený, J., Neef, J., Vojtěch, O., and Burda, H. (2008) "Magnetic alignment in grazing and resting cattle and deer," *Proceedings of the National Academy of Sciences*, 105 (36): 13451–13455.

Bravo, M. (1999) "Ethnographic navigation and the geographical gift" in Livingstone, D. N. and Withers, C. W. J. (eds) *Geography and enlightenment*, Chicago: University of Chicago Press, 199–235.

Burda, H., Begall, S., Červený, J., Neef, J. and Němec, P. (2009) "Extremely low-frequency electromagnetic fields disrupt magnetic alignment of ruminants," *PNAS*, 106 (14): 5708–5713.

Burda, H., Marhold, S., Westenberger, T., Wiltschko, W., and Wiltschko, R. (1990) "Magnetic compass orientation in the subterranean rodent Cryptomys hottentotus (Bathyergidae, Rodentia)," *Experientia*, 46 (5): 528–530.

Calleja, G. (2011) *In-game: from immersion to incorporation*, Cambridge, MA: MIT Press.

Caquard, S. (2013) "Cartography I: mapping narrative cartography," *Progress in Human Geography*, 37 (1): 135–144.

Caquard, S. (2014) "Cartography II: collective cartographies in the social media era," *Progress in Human Geography*, 38 (1): 141–150.

Cartwright, W. E. and Hunter, G. J. (2001) "Towards a methodology for the evaluation of multimedia geographical information products," *GeoInformatica*, 5 (3): 291–315.

Couldry, N. (2010) "Theorising media as practice" in Bräuchler, B. and Postill, J. (eds) *Theorising media and practice*, New York: Berghahn, 35–54.

Dodge, M., Perkins C., and Kitchin, R. (2009) "Mapping modes, methods and moments: a manifesto for map studies" in Dodge, M., Kitchin, R. and Perkins, C. (eds) *Rethinking maps*, London: Routledge, 220–243.

Döring, J. and Thielmann, T. (2009) "Mediengeographie: für eine Geomedienwissenschaft" in Döring, J. and Thielmann, T. (eds) *Mediengeographie: Theorie – Analyse – Diskussion*, Bielefeld: transcript, 9–64.

Gartner, G. (2009) "Web mapping 2.0" in Dodge, M., Kitchin, R. and Perkins, C. (eds) *Rethinking maps*, London: Routledge, 68–82.

Goodchild, M. F. (2007) "Citizens as sensors: the world of volunteered geography," *GeoJournal*, 69 (4): 211–221.

Goodchild, M. F. (2008) "What does Google Earth mean for the social sciences?" in Dodge, M., McDerby, M. and Turner, M. (eds) *Geographic visualization*, Chichester: Wiley, 11–23.

Gordon, E. (2009) "The metageography of the internet" in: Döring, J. and Thielmann, T. (eds) *Mediengeographie: Theorie – Analyse – Diskussion*, Bielefeld: transcript, 397–411.

Haraway, D. (1988) "Situated knowledges: the science question in feminism and the privilege of partial perspective," *Feminist Studies*, 14 (3): 575–599.

Hardt, M. (1999) "Affective labor," *boundary*, 26 (2): 89–100.

Hind, S. and Lammes, S. (2016) "Digital mapping as double-tap: cartographic modes, calculations and failures," *Global Discourse*, 6 (1–2): 79–97.

Kingsbury, P. and Jones III, J. P. (2009) "Walter Benjamin's Dionysian adventures on Google Earth," *Geoforum*, 40 (4). 502–513.

Klatzky, R. L. (1998) "Allocentric and egocentric spatial representations: definitions, distinctions, and interconnections" in: Freksa, C., Habel, C. and Wender, K. F. (eds) *Spatial cognition: an interdisciplinary approach to representing and processing spatial knowledge*, Berlin: Springer, 1–17.

Korzybski, A. (2005) "What I believe," *ETC: A Review of General Semantics*, 62 (1): 17–27.

Latour, B. (1986) "Visualization and cognition: thinking with eyes and hands," *Knowledge and society: studies in the sociology of culture past and present*, 6: 1–40.

Latour, B. (1987) *Science in action: how to follow scientists and engineers through society*, Cambridge: Cambridge University Press.

Latour, B. (1999) *Pandora's hope: essays on the reality of science studies*, Cambridge, MA: Harvard University Press.

Latour, B. (2013) *An inquiry into modes of existence*, Cambridge, MA: Harvard University Press.

Law, J. (1986) "On the methods of long distance control: vessels, navigation, and the Portuguese route to India" in Law, J. (ed.) *Power, action and belief: a new sociology of knowledge?*, London: Routledge & Kegan Paul, 234–263.

Lombard, M. and Ditton, T. (1997) "At the heart of it all: the concept of presence," *Journal of Computer–Mediated Communication*, 3 (2). https://doi.org/10.1111/j.1083-6101.1997. tb00072.x.

McHaffie, P. (1995) "Manufacturing metaphors: public cartography, the market, and democracy" in Pickles, J. (ed.) *Ground truth: the social implications of geographic information systems*, New York: Guilford Press, 113–129.

November, V., Camacho-Hübner, E., and Latour, B. (2010) "Entering a risky territory: space in the age of digital navigation," *Environment and Planning D: Society and Space*, 28 (4): 581–599.

Pariser, E. (2011) *The filter bubble: what the internet is hiding from you*, New York: Penguin Press.

Perkins, C. (2008) "Cultures of map use," *The Cartographic Journal*, 45 (2): 150–158.

Petchenik, B. B. (1983) "A map maker's perspective on map design research" in Fraser Taylor, D. R. (ed.) *Graphic communication and design in contemporary cartography*, Chichester: Wiley, 37–68.

Peuquet, D. and Kraak, M.-J. (2002) "Geobrowsing: creative thinking and knowledge discovery using geographic visualization," *Information Visualization*, 1 (1): 80–91.

Pickering, A. (1995) *The mangle of practice: time, agency, and science*, Chicago: University of Chicago Press.

Pickles, J. (ed.) (1995) *Ground truth: the social implications of geographical information systems*, New York: Guilford Press.

Pickles, J. (2004) *A history of spaces: cartographic reason, mapping and the geo-coded world*, London: Routledge.

Pinch, T. J. and Bijker, W. E. (1987) "The social construction of facts and artifacts: or how the sociology of science and the sociology of technology might benefit each other" in Bijker, W. E., Hughes, T. P., and Pinch, T. J. (eds) *The social construction of technological systems: new directions in the sociology and history of technology*, Cambridge, MA: MIT Press, 17–50.

Schäfer, M. T. (2011) *Bastard culture! How user participation transforms cultural production*, Amsterdam: Amsterdam University Press.

Scharl, A. (2007) "Towards the geospatial web: media platforms for managing geotagged knowledge repositories" in Scharl, A. and Tochtermann, K. (eds) *The geospatial web*, London: Springer, 3–14.

Schatzki, T. R. (2006) "Introduction: practice theory" in Schatzki, T. R., Knorr-Cetina, K., and von Savigny, E. (eds) *The practice turn in contemporary theory*, London: Routledge, 10–23.

Schüttpelz, E. (2009) "Die medientechnische Überlegenheit des Westens" in Döring, J. and Thielmann, T. (eds) *Mediengeographie: Theorie – Analyse – Diskussion*, Bielefeld: transcript, 67–110.

Slater, M. and Wilbur, S. (1997) "A framework for immersive virtual environments (FIVE): speculations on the role of presence in virtual environments," *Presence*, 6 (6): 603–616.

Star, S. L. and Griesemer, J. R. (1989) "Institutional ecology, 'translations' and boundary objects: amateurs and professionals in Berkeley's Museum of Vertebrate Zoology, 1907–39," *Social Studies of Science*, 19 (3): 387–420.

Thielmann, T. (2007) "You have reached your destination! Position, positioning and superpositioning of space through car navigation systems," *Social Geography* 2 (1): 63–75.

Thielmann, T., van der Velden, L., Fischer, F., and Vogler, R. (2012) "Dwelling in the web: toward a googlization of space," HIIG Discussion Paper Series No. 2012–03, available online at: http://papers.ssrn.com/sol3/papers.cfm?abstract_id=2151949 (accessed September 20, 2016).

Thrift, N. (2006) "Re-inventing invention: new tendencies in capitalist commodification," *Economy and Society*, 35 (2): 279–306.

Turner, A. J. (2006) *Introduction to neogeography*, Sebastopol: O'Reilly Short cuts.

Verhoeff, N. (2012) *Mobile screens: the visual regime of navigation*, Amsterdam: Amsterdam University Press.

Werlen, B. (2008) "Körper, Raum und mediale Repräsentation" in Döring, J. and Thiel-mann, T. (eds) *Spatial turn: Das Raumparadigma in den Kultur– und Sozialwissen-schaften*, Bielefeld: transcript, 365–392.

Zook, M. and Graham, M. (2007) "The creative reconstruction of the internet: Google and the privatization of cyberspace and DigiPlace," *Geoforum* 38 (6): 1322–1343.

Part III

Politics and inequalities

Studies on structural conditions and resources as the given context of acting with digital media and technology constitute a major part of scientific geographies of digital culture. These structural conditions frame practices and discourses of integration and exclusion, and they reflect the material and spatial uneven distribution of access to and knowledge about digital infrastructures. For example, new divisions of labor, classifications of users, or different levels of wellbeing (or health quality) emerge in the course of digitalization. This section primarily contains contributions that focus on digital culture's structural and physical relations with political and societal issues.

The striking contrasts in the social distribution of access to information via the internet are introduced and discussed by Barney Warf in his text on "Digital divides in the twenty-first century United States." Several of the common factors that usually determine social inequality such as race, age, gender, or the place of living are reflected in the inequalities that arise in the course of highly uneven developments of digitalization. While the promise of the digital transformation of society was to provide equal access and, thus, enhance democratic political and unconfined social participation, empirical facts and findings suggest that traditional inequalities persist within the realms of digital culture.

Fabio Contel's "The diffusion of information technologies in the Brazilian banking system and the indebtedness of low-income population" addresses the digitalization of the banking sector. For a long time, a large proportion of Brazilians were not part of the banking realm and did not hold an individual bank account; thus, they could not use many banking services, e.g., participate in electronic money transfer or receive a loan. Digitalization changed this asymmetrical demographic representation within the financial economy in several steps. It started with the rise of small, rather informal, agencies that exceeded the reach of classic bank branches in providing simple services for people that were not customers of a particular bank. Currently, digital media, communication, and their spread across many classes and milieus open up new possibilities, provide cheap and simple services, but they also cause new problems such as the increase of debt among low-income groups.

The role of digital culture is not limited to immaterial phenomena such as communication, cultural meanings, and its virtual spaces of symbols and imaginations. As Annika Richterich shows in her chapter on "Digital health mapping," physical phenomena such as diseases and health issues are surveyed and represented by digital social media platforms. They combine the medical and technological expertise of the organizations that run them with the potential of crowdsourced volunteered information and observations by lay people as their main contributors. The geographical map (as the main form in which the uploaded and processed data is represented) provides an example for the practical relevance of digital media and digital mapping techniques. At the same time, it becomes clear that these techniques are not only a matter of simple and correct representation—they are also deliberately (and responsibly) structured geospatial constructions.

8 Digital divides in the twenty-first century United States

Barney Warf

Information technologies have given more people more access to more information than at any time in human history. The ability to acquire, process, and
transmit information has become central to individual happiness, success, and
social mobility, so much so that simple dichotomies like offline/online fail to do
justice to the ways in which the real and virtual worlds are intertwined. Although
the internet has become almost ubiquitous in Western societies, significant pools
of people are still excluded from cyberspace, notably the elderly, poor, uneducated, disabled, and many ethnic minorities. This chapter confronts the thorny
matter of digital divides, the profound social and spatial discrepancies in access
to digital technologies that are found in many areas. Even in the most hard-wired
of cities, pockets of persistently offline people exist. While people who use the
internet regularly swim in deep oceans of information, people without internet
access are profoundly handicapped in a variety of ways, including the ability to
search for jobs, acquire skills, engage in electronic banking and bills payment,
apply for government forms and credit cards, and use cyberspace for informational, educational, or recreational purposes. Using data from the United States
for 2000–2015, the chapter explores topics such as the "racial ravine," or relatively low internet usage among non-whites; the roles of schools and public
libraries in overcoming the digital divide; the broadband digital divide; and the
implications of the mobile internet, i.e., access via "smartphones." Throughout,
it seeks to demonstrate that social and spatial inequalities are reinscribed within
cyberspace.

By now, digital reality and everyday life for hundreds of millions of people have
become so thoroughly fused that it is difficult to disentangle them. The internet is
used for so many purposes that life without it is simply inconceivable for vast
numbers of people. From email to online shopping and banking to airline and hotel
reservations, playing multi-player video games to chat rooms, Voice over Internet
Protocol telephony to distance education, streaming music and television shows to
blogs, YouTube to simply "Googling" information, the internet has emerged as
much more than a luxury to become a necessity for vast swaths of the population
in the economically developed world. In this context, simple dichotomies such as
"offline" and "online" fail to do justice to the diverse ways in which the "real" and
virtual worlds for hundreds of millions are interpenetrated.

Yet for many others—typically the poor, the elderly, the undereducated, ethnic minorities—the internet remains a distant, ambiguous world. Denied regular access to cyberspace by the technical skills necessary to log on, the funds required to purchase a computer, or public policies that assume their needs will be addressed by the market, information have-nots living in the economically advanced world are deprived of many of the benefits that cyberspace could offer them. While those with regular and reliable access to the internet often drown in a surplus of information (including vast amounts of spam), those with limited access have difficulty comprehending the savings in time and money it allows and the convenience and entertainment value it offers. As the uses and applications of the internet have multiplied rapidly, the opportunity costs sustained by those without access rise accordingly. At precisely the historical moment that contemporary capitalism has come to rely upon digital technologies to an unprecedented extent (Malecki and Moriset 2008; Zook 2005), large pools of the economically disenfranchised are shut off from cyberspace. In a society increasingly shaped by digital technologies, lack of access to cyberspace becomes ever more detrimental to social mobility and relationships, rendering those excluded from the internet more vulnerable than ever before (Graham 2002).

In June 2016, roughly 3.6 billion people, or 49 percent of the planet, used the internet on a regular basis (Internetworldstats 2016). Internet penetration rates, however, vary widely across the world. The United States continues its long-standing position as one of the world's societies with abundant access to the internet. Although internet penetration rates in the U.S. in 2015 (87 percent according to the census, 84 percent according to the Pew Charitable Trust, 2016) are not as high as Scandinavian nations (where they exceed 98 percent), they remain higher than many other urbanized, industrialized countries, and Americans as a whole still constitute one of the largest national blocks of internet users on the planet. Despite this prominence, there exist important discrepancies in internet access within the U.S. in terms of age, income and class, ethnicity, and location. As a slew of books has demonstrated, the digital divide is real, rapidly changing, complex, difficult to measure, and even more difficult to overcome (Andreasson 2015; Kuttan and Peters 2003; Stevens 2006; Van Dijk 2005; Warschauer 2003). While some decry the divide as a catastrophe, others deny its very existence.

The divide can be assessed in several ways. Simple access/no-access dichotomies do not capture its various aspects. The divide varies considerably by class, age, and ethnicity, as well as among places, notably between urban and rural areas. Access to broadband has become an important part. Inequalities in connection speed, reliability, and digital skills construct a multitude of divides. For many Americans, the most important number defining their online experience is the megabits-per-second (Mbps) available through their technological platform. Differential distributions of human capital (i.e., the ability to access the internet) are another. Indeed, the digital divide is so multidimensional that it cannot be reduced to dichotomous measurements; rather, it should be seen as a continuum measured across a variety of variables.

This chapter examines the changing social and spatial differentials in access to the internet in the U.S. in the period between 2000 and 2015. "Access," of course, is a nebulous term that exhibits different meanings (e.g., access at home, school, or work); perhaps the multiplicity of meanings is optimal for conveying the complexity of the digital divide, which does not lend itself easily to simple dichotomies (DiMaggio et al. 2001). Equally important as access is what users *do* with the internet, for simple access does not automatically lead one to become an internet user. Although the ability to gain access to the internet at work, home, school, or public libraries is widespread, employing cyberspace to gain meaningful information is another story. For many users, the internet will remain primarily a toy. Thus, assessments of internet usage must take into account the perspectives of the various populations that deploy it (or not) for their own means.

First, the chapter summarizes the various economic and political forces that have altered patterns of internet access in the U.S. Central to understanding the digital divide is the rapid growth in computer and internet usage among many social groups: the divide, such as it is, is never frozen in time or space, but a fluid, malleable entity that constantly shifts in size, composition, meaning, and implications. Second, it charts the growth in the absolute and relative numbers of different groups of American internet users in terms of their access at home and at work from 2000 to 2015. Third, it focuses on the critical issue of broadband delivery, which has reproduced older patterns of inequality. The conclusion explores the changing meanings of the American digital divide in an age in which access has become widespread, internet usage is of unparalleled importance, market imperatives dominate, and the consequences of not getting online are ever more profound. Throughout, it argues that the divide is not simply "digital," but profoundly social, political, and spatial.

Forces changing and perpetuating the U.S. digital divide

Several factors have conspired to dramatically accelerate internet access and usage in the U.S. among different social groups, including four major sets of forces: the declining costs of personal computers (PCs); public policies aimed at closing the digital divide; the deregulation and changing industrial structure of the telecommunications industry; and changing accessibility patterns in public schools and libraries.

Declining personal computer costs

The continued decline in the price of PCs looms as a major factor in expanding access to the internet. Following Moore's Law, which speculates the cost of computers falls in half roughly every 1.5 years, PCs have become increasingly ubiquitous across the U.S. Indeed, relatively fast, low-end machines are readily available for less than US$600 in numerous retail outlets. Almost 80 percent of Americans use a PC once or more per week either at work or at home, the vast bulk of which are networked. As the value of a network rises proportional to the square of the number of users (Zipf 1946), the internet and the PC made each other increasingly powerful and attractive. Simultaneously, the rise in user-friendly graphics interfaces such as the Netscape web browser greatly facilitated internet access for the parts of the population lacking in sophisticated computer skills. Moreover, as the number of applications of the internet has grown, the hours of usage have steadily increased to more than nine hours per week.

Changing public policies and structure of the telecommunications industry

Changes in public policy—including the deregulated environment unleashed by the 1996 Telecommunications Act—also shape the contours of the U.S. digital divide. Among other things, the Act was designed to encourage competition in high-cost rural areas and deliver the same access to cyberspace as found in cities. The Clinton Administration actively sought to reduce the digital divide by inserting the E-rate program (officially the Schools and Libraries Program of the Universal Service Fund) into the Act, which generated US$2.25 billion to provide discounts to telecommunications services ranging from 20 to 90 percent for low-income schools. E-rate was credited with raising the proportion of schools with internet access from 14 percent in 1996 to 95 percent in 2005. However, it did not provide funding for hardware, software, technological training, or access to broadband services.

The administration of George W. Bush was reluctant to intervene in what it deemed market imperatives; a policy of "technology neutrality" designed to avoid "market distortions." In practice, this strategy accentuated discrepancies in internet access. Typically, the Bush Administration argued that the divide has diminished to the point of irrelevance. In 2003, the Administration ended funding for two institutions central to previous efforts to minimize the divide, the Technology Opportunities Program in the Department of Commerce and the Community Technology Center initiative in the Department of Education. Instead of promoting universal access, the Administration excused cable television and telephone companies from this public service obligation.

The Obama Administration emphasized broadband and wireless services, as well as net neutrality, a paradigm that uncouples internet service providers (ISPs) from the content sent over their networks, which disallows price discrimination in favor of wealthier customers. Taken together, these policies encouraged telecommunications providers to offer services on a "pay per" basis, allowing them to "cherry pick" the most profitable customers and abandon those without significant purchasing power.

In the private sector, waves of corporate consolidation reshaped the landscape of telecommunications ownership and, correspondingly, the abilities of different social groups to get online. The market structure of telecommunications services has undergone a sustained transformation, including steady oligopolization. Like many sectors of telecommunications, ISPs were heavily affected by a wave of mergers and acquisitions, particularly after the 1996 Telecommunications Act, which greatly facilitated the process. Most ISPs lease capacity on fiber optics lines from telecommunications companies, many of which are publicly regulated, in contrast to the unregulated state of the internet itself. The privatization of the internet, which began in 1993 with the National Science Foundation's (NSF) transfer of the system's management to a consortium of private firms, increasingly brought it gradually into conformity with the dictates of the market. The resulting pattern of service provision became steadily restructured by corporate ISPs in partnership with backbone providers (e.g., AT&T, Worldcom, and Sprint), generating a geography centered largely on large metropolitan areas, whose concentrations of affluent users generate economies of scale that lead to the highest rates of profit (Warf 2003, 2013).

Urban-rural differentials

As in many countries, there are important differences between urban and rural areas in internet access. For example, 78 percent of rural Americans use the internet, compared to 85 percent of urban Americans (Perrin and Duggan 2015). The Federal Communication Commission (FCC 2014) defines broadband as "internet access that is always on" and provides service at speeds of "25 Mbps for downloads and 3 Mbps for uploads." However, this definition can vary based on the interests involved. A group of U.S. Senators recently criticized the paltry standards of the U.S. Department of Agriculture's Community Connect program, which allows providers with speeds of just 4 Mbps to claim that they offer "broadband." The National Telecommunications Industry Association further dilutes the meaning of "broadband," including in its definition services that provide speeds as low as 768 kilobits-per second (Kbps) downstream and 200 Kbps upstream (Mack and Grubesic 2014).

Overall, half of rural Americans lack access to service that meets the FCC's standard (Wheeler 2014). Specifically, 72 percent of rural dwellers are unable to access the internet at speeds greater than 3 Mbps. Slow and antiquated dial-up connections are disproportionately found in rural areas, while fiber optic lines are more likely to serve residents of high-density regions due to the economies of scale available there. Only 54.6 percent of rural Americans are able to access the internet at download speeds greater than 25 Mbps, compared to 94 percent of urban Americans. Intermediate technologies such as digital subscriber lines (DSLs), cable modems, and fixed wireless have also been deployed at disproportionately slow rates in low-density areas (Grubesic and Murray 2004). Even levels of mobile wireless coverage are often negatively associated with rurality (Riddlesden and Singleton 2014).

Access via public schools and libraries

Schools remain perhaps the most important arena in which the digital divide is manifested and reproduced (Monroe 2004). Given the lack of a national school system and reliance upon local property taxes as the primary means of funding public education, the U.S. school system tends to reinforce and deepen social inequalities rather than reduce them (Kozol 2005). In an age in which the acquisition of skills to participate in advanced producer services is key to upward social mobility, this issue assumes special importance. Inequalities in school funding are mirrored in the prevalence of the internet in public classrooms (Becker 2000): while 99 percent of schools offer children access to networked PCs in one way or another, these rates vary significantly in terms of quality of access. "[S]tudents with internet-connected computers in the classroom, as opposed to a central location like a lab or library, show greater improvement in basic skills" (Kaiser Foundation 2004). A similar situation applies to broadband access: roughly 41 percent of rural schools lack access to high-speed fiber connections (Wheeler 2014). Not surprisingly, the digital divide in schools has strongly racialized overtones: white students are much more likely than minorities to use the internet in the classroom or school library (U.S. Department of Education 2006). Of course, access to the mobile internet via smartphones mitigates these discrepancies. Nonetheless, some low-income students must use WiFi services at McDonald's to do their homework (Kang 2016).

Simple access to PCs at school is a poor measure of the extent of the digital divide: low-income students are less likely to have them at home or to possess the requisite technical skills necessary to install, maintain, and navigate such machines. Students with access at home are more likely to be enrolled, to graduate from high school, to go to university, and to have better grades than those who do not (Barzilai-Nahon 2006; Fairlie 2005). While 98 percent of all U.S. children aged eight to 18 years have "ever" gone online, regular, reliable, and rapid access to the internet with social and technical support, in a comfortable, non-distracting environment, remains stratified by ethnicity and family income.

After home and school, public libraries are the third-most common point of internet access, especially for lower-income minorities. Public libraries have been at the forefront of efforts to reduce the digital divide, and about 99.1 percent of all U.S. libraries offer free internet use. In many communities, libraries are the *only* free access to the internet. However, libraries have limited space and operating hours, often lack high-speed connections, and frequently find their limited information technology budgets strained by growing numbers of people such as the unemployed seeking to use their resources for job seeking, students using them for school, work, or others hoping to acquire computer skills (Walsh 2007). In 2007, the Bill and Melinda Gates Foundation announced a multi-year technology grant program for public libraries as part of its effort to combat the digital divide (Bill & Melinda Gates Foundation 2004).

The changing profile of the U.S. digital divide

Throughout the 2000–2015 period, growth in internet use among various socio-demographic groups was rapid, often spectacular (Table 8.1). Average internet penetration rates—including access at home, work, or school—rose by roughly

Table 8.1 Growth in adult U.S. internet users, 2000–2015

	% online in		% growth	% with broadband
	2000	2015		
Age				
18–29	70	96	37	81
30–49	61	93	52	77
50–64	46	81	76	68
65+	14	58	314	47
Total	52	84	61	70
Gender				
Men	54	84	55	70
Women	50	84	68	70
Ethnicity				
White	53	85	60	74
Black	38	78	105	62
Latino/Hispanic	46	78	69	56
Asian	72	97	35	n.a.
Education				
<High school	19	66	247	28
High school graduate	40	76	90	58
Some college	67	90	34	80
College graduate	78	95	22	90
Household income				
<US$30,000	34	74	117	52
US$30,000–49,000	58	85	46	71
US$50,000–75,000	72	95	32	85
>US$75,000	81	95	17	91
Community type				
Rural	42	78	86	60
Suburban	56	85	52	74
Urban	53	85	60	70

Sources: Pew Charitable Trust 2016, 2017a.

50 percent, from 52 percent to 84 percent; by 2015, 281 million Americans were using the internet regularly. Thus the innovation, the most rapidly diffused technology in world history, went from a tool or toy of a minority to an essential implement used by the vast majority. Every social group, as differentiated by age, gender, race/ethnicity, educational level, or household income, experienced marked gains. To the extent that the digital divide persists in the U.S. (and other economically advanced countries), it must be understood within the context of this sustained and rapid increase in the number of users and proportion of the population.

This growth, however, did not occur at identical rates among all social categories. Take, for instance, age, as measured in four broad categories. The young (i.e., under 30 years of age) steadily exhibited the highest internet penetration rates, reaching 96 percent in 2015. For many children who grow up surrounded by digital technologies, the internet is hardly mysterious. In contrast, in both benchmark years, the elderly (ages 65 and over) experienced the lowest rates of internet usage (a mere 14 percent in 2000 vs. 58 percent in 2015), as well as the slowest rate of increase in users. Many elderly people find new technologies to be difficult or intimidating, do not appreciate the potential benefits, are easily frustrated by their lack of technical skills, and are comfortably ensconced in their pre-internet lives. The digital divide, therefore, is closely wrapped up with generational differences, and the views and preferences of different groups of users are vital to understanding their willingness (or not) to participate in cyberspace.

Notably, gender differences in internet usage, which were small in 2000, declined steadily throughout this period, so that by 2015 they had disappeared. Despite its early reputation as an exclusive haven of masculinity, the internet in fact has been harnessed by increasing numbers of women. Gender differentials in access reflect both the lower socio-economic status of women relative to men as well as sexist cultural attitudes toward science and technology. The disappearance of the gender gap speaks to the increasing familiarity with digital technologies among many women, particularly the young and well educated, who are often employed in producer services in which minimal computer skills are an essential prerequisite. Moreover, enrollment rates in American universities for women have consistently surpassed those for men (Lopez and Gonzalez-Barrera 2014).

One dimension of the U.S. digital divide that has drawn the most serious scrutiny concerns racial or ethnic differences. Given the profound inequalities in U.S. society in terms of income, educational opportunities, and employment that exist between whites and ethnic minorities, it is not surprising that this gap is manifested in terms of access to cyberspace, i.e., much of the racial ravine in digital access is due to income discrepancies (Fairlie 2005). In 2000, internet access rates were highest for Asian-Americans, a relatively well-educated group. That for whites remained well above those for Latinos/Hispanics (53 vs. 38 percent) and almost double that of Blacks or African-Americans (46 percent). However, income alone does not explain the totality of the digital divide, as internet use and adoption are intertwined with cultural preferences of different ethnic populations.

There are signs, however, that this dimension of the digital divide is slowly diminishing. By 2015, rates of internet access among ethnic groups had steadily converged: white usage rates remained higher but not by as large of a margin. It should be noted that there are important differences within minority populations, however. Among African-Americans, internet usage tends to be concentrated among the young and the college-educated. Likewise, the Latino population is far from homogeneous, and significant discrepancies in internet access and usage remain among various sub-groups; usage rates tend to be much higher among bilingual Latinos than those who speak only Spanish. Indeed, among English-dominant Latinos, internet usage rates are identical to whites.

Among Native Americans, a sharp bifurcation exists between those living in urban areas, whose rates of access and usage mirror the country as a whole and those living on reservations, the proportion of whom using the internet falls well below the national mean. Some Native Americans view the internet as another tool of cultural assimilation, the latest in a long, sad history. While some universities (e.g., Northern Arizona University) offer free internet services to reservations, in general such places are politically inconsequential and unable to confront telecommunications companies (e.g., over rights of way issues). The Bill and Melinda Gates Foundation's Native American Access to Technology Program has successfully worked with tribes in the Four Corners area of Utah, Colorado, Arizona, and New Mexico to increase access to digital information resources while preserving local heritages.

Persistently underlying the digital divide in the United States are vast socioeconomic differences, particularly those in education and household income, which effectively serve as markers of class. Although populations at all of four broad educational levels (less than high school, high-school graduate, some college, college graduate) exhibited gains in internet access, profound differences remain. Among college-educated Americans, internet usage is almost universal (95 percent); however, only two-thirds of those lacking a high-school degree had access. There was, however, explosive growth among the least educated, whose internet penetration rates during 2000–2015 more than tripled.

Similarly, income remains a useful measure of who has access and who does not, particularly at home. In 2000 roughly 81 percent of upper-income households (over US$75,000 annually) used the internet; by 2015, this share had risen to 95 percent (Table 8.1). Rapid growth rates also occurred among those of more modest means, and today the vast majority of even poor households (earning less than US$30,000 annually) are users. Thus, as with race/ethnicity and educational level, absolute discrepancies persist but relative differences declined as internet usage rates advanced most rapidly among those with hitherto the least access.

It should be emphasized that American non-users of the internet are a surprisingly diverse bunch. They consist disproportionately of poorly educated women and minorities, and live in rural areas. One-quarter of non-users have not completed high school, compared to 5 percent of internet users. Non-users are much more likely than users to be retired or unemployed. Another 17 percent consist of "internet drop-outs," who typically became frustrated by their hardware, software, or service provider. Yet others consist of the disabled, particularly those who suffered severe strokes, and the blind, who lack or cannot afford Braille interfaces. Finally, a small but stubborn core of avowed non-users remain excluded from cyberspace not by income or education, but simply out of personal choice, saying they simply do not need the internet. While some cite the cost of computers and online service access, or say that it is simply too complicated, others cite fears of internet pornography, credit card fraud, or identity theft. Roughly one-quarter of this group struggles with literacy in their everyday lives, and this group is less likely than other non-users to know of public internet access points.

It is worth emphasizing that across various socio-economic categories, discrepancies in internet usage have dropped sharply. Thus, the higher rates enjoyed by the wealthy, whites, and the well educated remain, but have diminished markedly. The most rapid growth in usage has occurred among the most disenfranchised groups, while growth in usage among the advantaged has tapered off, in what are sure to be nearly saturated markets. In short, the continuing diffusion of internet usage has trimmed digital divides across the board.

The digital divide in the broadband arena

Broadband has become increasingly central to internet access. As web-based material has become increasingly graphics-based, involving the transmission of large, data-intensive files (e.g., photographs, video), broadband access has become correspondingly more important. Broadband applications include digital television, business-to-business linkages, internet gaming, telemedicine, video-conferencing, and internet telephony. With large, graphics-intensive files at the heart of most internet uses today (e.g., downloading forms, reading online newspapers, streaming video such as YouTube and Netflix), broadband has become increasingly imperative for efficient web browsing. Broadband is also reflective and a driving force behind the phenomenon of digital convergence, the blurring of boundaries that traditionally separated industries such as telephone, cable television, and computers, allowing the generation of significant economies of scope and scale.

Broadband technology has existed since the 1950s, but its widespread use was not economically feasible until the deployment of large quantities of fiber optics cable in the 1990s allowed vast amounts of data to be transferred at high speeds (up to 2.4 gigabytes per second). While trunk fiber lines stretch across the country and the world, many local loops into homes and businesses still use relatively slow twisted pair copper wires, giving rise to the famous "last mile" problem, that is, the difficulty connecting to users' homes when the final segment consists of slow, often twisted copper cable, connections.

In passing the Telecommunications Act of 1996, the U.S. Congress directed the FCC to encourage the growth of advanced telecommunications technologies, a directive that stimulated providers to offer fiber optic services directly into homes and businesses. Several technologies meet FCC standards for advanced services, which specify a very low minimum baud rate of 200 kbps, thus disqualifying ISDN connections, which operate at 144 kbps. Of the various options, DSL provided by cable television companies are the most popular; two-thirds of American households have cable television and many couple internet and television service into one integrated package. In addition, Asymmetric Digital Subscriber Lines (ADSL) include a suite of broadband technologies provided by local telephone companies that operate on twisted copper pairs and provide an "always on" internet connection, unlike traditional modems. Broadband adoption has also been encouraged by steadily declining prices in this market. As a result, the number of broadband lines jumped markedly.

In 2015, roughly 70 percent of the U.S. population used broadband technologies at home, the growth of which reduced dial-up services to marginal status. Non-users of broadband typically cite the expense or lack of availability in their local area as their reasons. Broadband accessibility tends to be most prevalent among the young, whites, the well educated, and rises monotonically with household income (Table 8.1), reflecting in many respects the same differentials that have accompanied dial-up internet since its inception. The elderly remain infrequent users of this mode of access, which was delivered only to 47 percent of those over age 65. Ethnic disparities remain in place with more users being white than Latino or African-American. Nonetheless, income and educational level remain the prime determinants of who has access to broadband and who does not. Such social differentials are accompanied by spatial ones. While 74 percent of suburban residents use broadband, as do 70 percent of urbanites, only 60 percent of rural denizens do so; however, growth rates were higher in rural than urban areas, indicating this discrepancy may decline in the future. Broadband technologies have been slow to reach rural America: whereas 86 percent or residents in cities with more than 100,000 residents have access to DSL, relatively few in towns with less than 10,000 people do so (Mack and Grubesic 2014; Riddlesden and Singleton 2014). Thus, there are strong reasons to believe that far from eliminating the digital divide, broadband reproduces it, gives it new form, and in some cases, accentuates it.

Despite its rapid growth, the proportion of broadband users in the U.S. is relatively low compared to most of the economically developed world; indeed, under the Bush Administration, the U.S. slipped internationally from fourth in 2001 to 15th in 2010 in terms of access to broadband services, and Americans pay ten times as much per megabit over broadband as do their counterparts in South Korea and Japan. Critics allege that the FCC exaggerated the extent of broadband usage in the U.S. (by including delivery speeds as low as 200 kbps, four times the speed of modem) and not taking the problems of inadequate access and low competition sufficiently seriously (e.g., Turner 2005); for example, the FCC holds a ZIP code as having broadband service if it contains only one subscriber, without consideration of price or speed.

However, the rapid growth in wireless and mobile broadband services injects complexity into this view. In 2016, approximately 77 percent of Americans had smartphones, which have replaced desktop computers as the most common form of internet access (Table 8.2). Ethnic differences in access are negligible, including 77 percent of whites, 75 percent of Latinos, and 72 percent of African-Americans. Age differences in access to the mobile internet are pronounced: smartphones are almost ubiquitous among the young but decline steadily as users get older. Educational level is closely associated with increased frequency of access. Smartphone usage increases steadily with income, as one might expect. Finally, the urban-rural digital divide reappears in the mobile internet as well: only two-thirds of rural residents own them, whereas 77 percent of urbanites and 79 percent of suburbanites do, respectively.

Table 8.2 Percent of Americans with smartphone and mobile internet access in 2016

Ethnicity	
White	77
Black	73
Latino/Hispanic	75
Age	
18–29	92
30–49	88
50–64	74
65+	42
Education	
<High school	54
High school graduate	69
Some college	80
College graduate	89
Income	
<US$30,000	64
US$30,000–49,000	74
US$50,000–75,000	83
>75,000	93
Community type	
Rural	77
Suburban	79
Urban	67

Source: Pew Charitable Trust 2017b.

Discussion

Contrary to common utopian interpretations, cyberspace is shot through with relations of class, gender, ethnicity, and other social categories. When viewed in social terms, the interpenetration of the virtual and real worlds is mutually constitutive: discrepancies in access to the internet simultaneously mirror and augment inequalities in the world outside of cyberspace.

The digital divide in the U.S. must be viewed in terms of the rapid absolute and relative growth in the number of users that occurred in the late 1990s and early part of the 2000–2010 decade. Today, 281 million people, or 87 percent of the U.S. population, have access to the internet either at home or at work. Among those with occupations demanding a university education, internet usage is almost universal. As the size of the U.S. internet population has grown, it has steadily come to demographically resemble the country as a whole. Many of the most egregious dimensions of the digital divide have been mitigated, although they have scarcely disappeared. Gender differences, for example, which once loomed large, have largely evaporated as females became as proficient at using the web as males. While whites continue to enjoy higher rates of access than do minorities, this gap has declined as well; the racial ravine has given way to a more modest ethnic gulch. Education level remains a prime marker of who has access and uses the internet and who does not. That such differentials have declined in the face of the indifference of the George W. Bush Administration testifies to the falling prices of computer hardware, the diffusion of software skills among ever larger segments of the population, and the role played by schools and public libraries.

However, class differences—as expressed through different access rates for varying levels of education and household income—remain an important dimension of the American digital divide. Vast swaths of the population—largely minority, poorly educated, low in income, and often employed in the lowest rungs of the service sector—have little experience with the internet. For many, cyberspace appears as some dimly perceived horizon with few concrete advantages to offer. Ironically, it is precisely such pools of people who might benefit the most by having access to for example, information about employment opportunities, bus schedules, or comparative shopping capabilities that the internet affords. Lack of reliable access deprives the poor and uneducated of the possibility of participating as equals (Stevens 2006). Since low-income ethnic minorities comprise a disproportionate share of new entrants into the labor force, the lack of internet skills among such workers is also a matter of national competitiveness. It is only when the bottommost tiers of the social order have reliable access that the digital divide will disappear, if it ever does. Until then, the internet may amplify social inequalities as much as it reduces them.

Moreover, important geographic variations remain: it is no accident that the highest rates of internet access are to be found in states with relatively good public education systems (e.g., the U.S. Northern Midwest) and relatively high per capita incomes. Conversely, the lowest rates are evident in poorer, frequently Southern states that typically underinvest in public education systems. Thus, the spatial dimensions of the digital divide mirror the socio-economic ones; *where* users are located has as much to do with access as who they are, for the social and the spatial are hopelessly intertwined.

Even with enormous price declines in the cost of PCs, considerable portions of the low-income population do not have them at home. Use of a networked PC, of course, presupposes minimal technical skills, which the country's least educated segments almost universally lack. As Korupp and Szydlik (2005) emphasize, social and family context and human capital matter as much or more than does the simple presence of a PC. Thus, attempts to overcome the digital divide by extending the internet to the poorest, least educated portions of the country will encounter steeply diminishing returns; it is one thing to offer simple access, and quite another to teach the computer illiterate the basic skills necessary to navigate cyberspace and participate in the information economy. However, as a new generation of younger users, increasingly familiar with the internet, gradually replaces their less computer-oriented elders, much of the roughest contours of the digital divide may be ameliorated over time.

The contemporary frontier that speaks most accurately to the digital divide's evolving nature is the uneven social and spatial distribution of broadband services. Given that the bulk of internet applications are graphics-intensive, including web-based functionality, broadband has become increasingly essential to meaningful internet usage. Typically, given the deregulated climate of the telecommunications industry, providers seek to avoid low-income or rural areas (where low densities inhibit economies of scale) and "cherry pick" relatively affluent, densely populated urban ones. Thus, rural-urban differences in internet access—a topic woefully understudied in the academic literature—remain critical to understanding who has access and who does not.

Conclusion

The digital divide in the U.S. reflects the unique constellation of cultural, political, and economic forces that have long defined American society: its high degree of individualism; its faith in mythical free markets and distrust of state intervention; its tolerance of inequality; and the profoundly racialized nature that permeates differential access to social opportunities, including the internet. Unequal access to the internet reflects broader, growing inequalities generated by labor market polarization (including the loss of manufacturing jobs and the explosion of low-wage services), the growth of unearned income (particularly dividends), and a largely indifferent federal government.

What might be done to reduce the digital divide in the future? Three lines of action present themselves. First, universal service provisions, largely abandoned after the 1996 Telecommunications Act, should be reinstated as part of any federal government regulatory programs. Since the market for internet services is unlikely to provide access for low-income populations by itself, this type of policy stipulation lies at the core of any effective public program to reduce disparities in access. Second, subsidized partnerships between telecommunications companies and ISPs should address public schools and libraries in low-income neighborhoods, including a revival and expansion of the e-rate program, and focus not simply on the provision of computer hardware but also on the generation of human capital, i.e., the skills necessary to log on, navigate the internet, and employ it in substantively meaningful ways. Finally, aggressive efforts should be made to encourage broadband and mobile internet access, including subsidies to overcome the last mile problem in impoverished regions and the proliferation of wireless "hot spots." Given how entrenched inequality is in the United States, such measures will require substantial investments and lengths of time to be effective; what is clear is that without them, the digital divide will persist.

References

Andreasson, K. (ed.) (2015) *Digital divides: the new challenges and opportunities of e-inclusion*, London: Taylor and Francis.

Barzilai-Nahon, K. (2006) "Gaps and bits: conceptualizing measurements for digital divide/s," *The Information Society*, 22 (5): 269–278. http://dx.doi.org/10.1080/019722 40600903953.

Becker, H. J. (2000) "Who's wired and who's not: children's access to and use of computer technology," *The Future of Children*, 10 (2): 44–75.

Bill & Melinda Gates Foundation (2004) "Toward equality of access: the role of public libraries in addressing the digital divide," available online at: www.imls.gov/assets/1/ AssetManager/Equality.pdf (accessed July 21, 2017).

DiMaggio, P., Hargittai, E., Newman, W. and Robinson, J. (2001) "Social implications of the internet," *Annual Review of Sociology*, 27: 307–336. https://doi.org/10.1146/ annurev.soc.27.1.307.

Fairlie, R. (2005) "Are we really a nation online? Ethnic and racial disparities in access to technology and their consequences," available online at: www.civilrights.org/issues/ communication/digitaldivide.pdf (accessed July 21, 2017).

Federal Communications Commission (2014) *Types of broadband connections*, Washington, DC: Federal Communications Commission.

Graham, S. (2002) "Bridging urban digital divides? Urban polarisation and information and communication technologies (ICT)" *Urban Studies*, 39 (1): 33–56.

Grubesic, T. and Murray, A. (2004) "Waiting for broadband: local competition and the spatial distribution of advanced telecommunication services in the United States," *Growth and Change*, 35 (2): 139–165.

Internetworldstats (2016) "Usage and population statistics," available online at: www. internetworldstats.com (accessed May 6, 2017).

Kaiser Foundation (2004) "Children, the digital divide, and federal policy," available online at: https://kaiserfamilyfoundation.files.wordpress.com/2013/01/children-the-digital-divide-and-federal-policy-issue-brief.pdf (accessed July 20, 2017).

Kang, C. (2016) "Bridging a digital divide that leaves schoolchildren behind," available online at: www.nytimes.com/2016/02/23/technology/fcc-internet-access-school.html (accessed July 26, 2017).

Korupp, S. and Szydlik, M. (2005) "Causes and trends of the digital divide," *European Sociological Review*, 21 (4): 409–422. https://doi.org/10.1093/esr/jci030.

Kozol, J. (2005) *The shame of the nation*, New York: Random House.

Kuttan, A. and Peters, L. (2003) *From digital divide to digital opportunity*, Lanham: Scarecrow Press.

Lopez, M. and Gonzalez-Barrera, A. (2014) "Women's college enrollment gains leave men behind," available online at: www.pcwresearch.org/fact-tank/2014/03/06/womens-college-enrollment-gains-leave-men-behind/ (accessed July 22, 2017).

Mack, E. A. and Grubesic, T. H. (2014) "US broadband policy and the spatio-temporal evolution of broadband markets," *Regional Science Policy & Practice*, 6 (3): 291–308.

Malecki, E. and Moriset, B. (2008) *The digital economy: business organisation, production processes, and regional developments*, London: Routledge.

Monroe, B. (2004) *Crossing the digital divide: race, writing, and technology in the classroom*, New York: Teachers College Press.

Perrin, A. and Duggan, M. (2015) *Americans' internet access: 2000–2015*, Washington, DC: Pew Research Center.

Pew Charitable Trust (2016) "Americans' internet access: 2000–2015," available online at: www.pewinternet.org/2015/06/26/americans-internet-access-2000-2015/ (accessed July 27, 2017).

Pew Charitable Trust (2017a) "Internet/broadband fact sheet," available online at: www. pewinternet.org/fact-sheets/broadband-technology-fact-sheet/ (accessed July 27, 2017).

Pew Charitable Trust (2017b) "Mobile fact sheet," available online at: www.pewinternet. org/fact-sheet/mobile/ (accessed July 22, 2017).

Riddlesden, D. and Singleton, A. (2014) "Broadband speed equity: a new digital divide?," *Applied Geography*, 52: 25–33. https://doi.org/10.1016/j.apgeog.2014.04.008.

Stevens, D. (2006) *Inequality.com: money, power and the digital divide*, Oxford: Oneworld Publications.

Turner, S. (2005) "Broadband reality check: the FCC ignores America's digital divide," available online at: www.savetheinternet.com/sites/default/files/resources/broadband_report.pdf (accessed July 22, 2017).

U.S. Department of Education (2006) "Computer and internet use by students in 2003: statistical analysis report," available online at: https://nces.ed.gov/pubs2006/2006065. pdf (accessed July 22, 2017).

Van Dijk, J. (2005) *The deepening divide: inequality in the information society*, Thousand Oaks: Sage.

Walsh, T. (2007) "Libraries strained by internet use," available online at: https://gcn.com/ articles/2007/09/20/libraries-strained-by-internet-use.aspx?m=2 (accessed July 22, 2017).

Warf, B. (2003) "Mergers and acquisitions in the telecommunications industry," *Growth and Change*, 34 (3): 321–344.

Warf, B. (2013) "Contemporary digital divides in the United States," *Tijdschrift voor Economische en Sociale Geografie*, 104 (1): 1–17.

Warschauer, M. (2003) *Technology and social inclusion: rethinking the digital divide*, Cambridge, MA: MIT Press.

Wheeler, T. (2014) "Closing the digital divide in rural America," available online at: www.fcc.gov/news-events/blog/2014/11/20/closing-digital-divide-rural-america (accessed July 22, 2017).

Zipf, G. (1946) "Some determinants of the circulation of information," *American Journal of Psychology*, 59 (3): 401–421.

Zook, M. (2005) *The geography of the internet industry*, Oxford: Wiley-Blackwell.

9 The diffusion of information technologies in the Brazilian banking system and the indebtedness of low-income population

Fabio Betioli Contel

Introduction

This chapter aims to understand the recent restructuring of the Brazilian banking system from a geographic view. The goal is to analyze how this restructuring led to increased indebtedness of the low-income population in Brazil. Based on the literature and documents studied, it was possible to find causal links between the diffusion of a new bank topology (more intensively constituted with informational objects) and the rise of banking among the population, especially those with lower income, which previously had no direct relationship with the financial system.

Despite being mainly widespread in the contemporary period, the use of financial instruments by large companies dates from the emergence of the so-called "financial capitalism" since the late nineteenth century. As described by Nicolai Bukharin ([1915] 1969: 71), "the financial capital is certainly the most invasive form of capital, suffering from *horror vacui*—the need to fill any empty space it might find." The geographer Jean Labasse (1976: 25) argues that there is a "colonizing will" in banking networks, guiding financial services to places where it is possible to drain resources (taking deposits, selling financial products) and to irrigate areas (lending money).

More recently, Anthony Giddens (1991: 32) affirms that kinds of financial instruments (mainly money, bank deposits, credit/debit systems) may be regarded as "symbolic tokens"—disembedding mechanisms in the daily life of economic agents. Within the "expert systems" (this new paraphernalia of technical objects flooding our environment), coins are symbolic tokens. They are powerful instruments promoting the "displacement of social relations from local contexts of interaction, also restructuring these relations across indefinite spans of time-space" (ibid.: 28).

How should one analyze this disembedding caused by the spread of financial variables in contemporary Brazil? Which geographical explanations could be given to this phenomenon? How does information technology influence the diffusion of banking products and services? Which new types of facilities, technical equipment, and products were recently made available to the Brazilian population? How do these new elements interfere in the everyday life of this population, especially in terms of debt? These are some of the questions that this chapter seeks to answer or at least proposes some solid interpretations to clarify these phenomena.

The diffusion of the technical-scientific-informational milieu and the new banking topology in Brazil

The diffusion in Brazil of what Milton Santos (1994, 1996) called "technical-scientific-informational milieu" began in the 1950s, when a number of technical macro-systems were built that enhanced the process of urbanization and allowed a significant modernization of the national economy. Since the 1950s, the spread of modern highways, powerful energy production systems (mainly hydroelectric plants and its gigantic transmission lines), and also a set of telecommunications infrastructures have increased the transmission capacity and the speed of circulation of information on national territory (Dias 1995; Santos 1993; Santos and Silveira 2000). This new material basis, in turn, allowed the installation of modern facilities, technical systems, and new products offered by banks that enabled the financial system to expand toward places where the population carries out their daily lives.

Traditional forms of banking topology have been changing due to its blend with information technology. With the incorporation of *informational objects* into this topology that was fundamentally structured in more analogic technical systems-end bank branches, we move to a more diverse type of banking networks—with the incorporation of automated teller machines (ATMs), banking correspondents, and new electronic channels such as call centers, internet banking, and mobile banking. These new technical systems increase the influence of the financial rationality in various populations of Brazil and could also increase the indebtedness of the low-income population, as we shall see here.

A key novelty in the Brazilian banking topology seems to be the so-called "banking correspondents." The correspondents are "micro-agencies" hired by large commercial banks and other institutions of retail credit. They aim to provide simple banking services such as receiving government aid (e.g., pensions and retirement benefits), paying low-value bills, and withdrawing low amounts of cash, among other functions.

Banking correspondents started to spread in the country from August 25, 1999, with the Resolution 2640 by the Central Bank of Brazil (later supplemented by Resolution 2707 of March 30, 2000). This resolution authorized such micro-agencies to offer these simple services that were previously provided almost exclusively by main bank branches. Although correspondents cannot be considered as "pure" informational/automated systems—since they depend on an employee to operate it—their spread depended on computers and General Packet Radio Service (GPRS) to transmit data from the correspondents to the agencies that hired them.

Regarding their location, the correspondents have a very flexible geographical diffusion. Since they require simple micro-structures—a computer, an employee, a balcony, and a network—they could be installed in various establishments that already existed. Lottery houses and post offices are the most common locations where they can be found, but it is also possible to find them in bakeries, pharmacies, building material stores, small shops, etc.

Due to this flexibility in location and to their low maintenance costs, the correspondents spread rapidly in Brazil. In their first year of existence, 1999, they were already present in 1679 of the total 5565 of Brazilian municipalities. In 2002, they existed in all the municipalities, making banking presence in Brazil strongly ubiquitous (Contel 2011). This ubiquity was decisive to facilitate the access of low-income population to banking services in Brazil. In addition to this, three main characteristics of correspondents were also crucial to enable this access: (1) Correspondents are always located in pedestrian areas of large circulation, such as bus terminals, subways, trains, promenades, major junctions of streets, etc.; (2) Their entrance system is more simple than in standard agencies, since they do not have metal detectors, revolving doors, or armed guards—decreasing physical constraints for potential customers; (3) As they are located in commercial establishments or services which are familiar to the local population, cultural constraints to enter are also decreased (Contel 2011).

Due to these reasons, the number of correspondents increased quite rapidly; in 2004 there were 46,035 correspondents operating in Brazil compared to 2008 where there were already 108,074, until reaching the impressive number of 375,315 in 2013—the year in which there was the highest number of correspondents in the country (FEBRABAN 2015).

In addition to banking correspondents, there was also a spread of new service channels with intensive use of information technologies, called "electronic financial channels." These allow transactions between two or more users without the physical presence of a human being, such as those carried out by ATMs, internet banking networks, call centers, and receiving points of credit and debit cards (called POS or points of sale in Brazil). These channels gave a much larger reach to the previous topology of the Brazilian financial system, both socially and geographically.

Changes in these channels are relatively recent and have accelerated in the last 20 years. Electronic channels that actually contributed to the amplification of banking networks in Brazil would be: (1) ATMs; (2) POS; (3) electronic check; (4) internet banking; (5) central telephone answering; (6) wireless application assistant (WAP); and (7) electronic data interchange systems. All these channels are extremely intensive in information technology, also allowing banks to develop remote services for support and sale. We then pass to a closer analysis of these technical systems.[1]

Implemented in Brazil in the 1980s, ATMs were primarily used within banks as a way to divert the provision of services of traditional cashiers. The system of this electronic channel can be described as an "electronic device consisting of hardware and software owned by a financial institution; internal modem; cameras for tracking and remote monitoring; data transmission via WAN (Wide Area Network) or agency network" (Silva 2005: 90).

In addition to the primary function of cash withdrawals, the following operations can be performed in an ATM: (a) check balances; (b) manage financial investments (and have the results printed or shown in the screen); (c) transfer values between accounts or/and applications; (d) make deposits; (e) print check sheets; (f) pay taxes; etc. This set of functions allows ATMs to function as mini-agencies, as described by Carlos Franco da Silva (1999: 60). These functions can be operated by ATMs in a kiosk, in the proper banks, or in shops of various kinds. Before the advent of ATMs, such operations could only be carried out within bank branches and with the direct participation of a member of the bank staff.

The operations of cashiers involve complex technical systems that circulate information, data, and money. The handling of bank notes and documents in automatic terminals is enabled by a mechanical and electronic system that can issue checks, dispense money, receive deposits, among other typical operations of a bank. As ATMs greatly increase the productivity of banking services, they also reduce the need for staff to do these services. Such automation, "sharply affects employment levels in the banking sector and intensifies the work in the bank field," as put by Nise Jinkings (2002: 124).

Carlos Franco da Silva (1999) and Jackeline Silva (2005) draw attention to the new temporality and greater accessibility provided by an ATM network in urban centers. They highlight the fact that ATMs are located in strategic places allowing customers to have easy access to services available with additional hours beyond the functioning of standard branches. Melo et al. (2000: 50) emphasize the possibility of operating transactions from inside a car as a main cause for the spread of ATMs in Brazil (drive-through), as well as its flexibility in location and working hours.

Call centers are another important informational channel in this renewal of the banking topology. The diffusion of this channel is directly linked to the expansion of the telephone network in Brazil. Although it dates from before the banking automation in the 1980s, it saw a boost, especially in the 1990s, and today it is very present in the routine of financial institutions and commercial enterprises. Telephone transactions were common before but they were made with the intervention of an attendant, whereas, with the advent of call centers they could be carried out by automatic means. "Through electronic resources such as IVR (Interactive Voice Response), it became possible to map and treat the transaction request of the customer and forward it for speech recognition and processing" (Silva 2005: 86).

In terms of accessibility, the use of electronic service centers also amplifies the capillarity of the financial system. The incorporation of the telephone network reinforces the fact that almost the entire national territory is covered by banking networks, since the use of telephones (both mobile and fixed) is highly widespread in Brazilian territory. Despite the importance of call centers for banks, their use has recently decreased, as explained later in this text.

Another result of this diffusion of a more informatized material basis is internet and mobile banking, which until the last decade was considered "the most popular electronic channel used for transactions" (Silva 2005: 84). Besides its high capillarity, the platform of the World Wide Web significantly reduces the transaction costs of products and services offered by banks. In addition to the use of internet banking, it is also increasingly common in Brazil to use smart-phones as a tool for carrying out banking transactions. According to a study carried out by the Brazilian Federation of Banks (FEBRABAN), the higher speed and function of these gadgets (compared with agencies and ATM) has boosted their use for banking purposes. FEBRABAN's study affirms that this usage increase resulted in a rise in the number of transactions, particularly the non-financial ones (mainly consultations of bank balances):

> These channels brought so much convenience that not only did they make clients leave the physical visits to other channels, but also encouraged them to become more active in their transactional relationships with banks. As a result of these factors, we note that users of Internet and Mobile Banking perform more transactions than those without access to virtual channels.
>
> (FEBRABAN 2014: 29)

In general, technical systems such as the World Wide Web and smartphones have allowed financial service networks to spread banking presence in geographic space. This diffusion is much stronger in urban centers, since they have larger densities of population and more public and private companies. To better understand the diffusion of such systems, we should highlight four main factors that are at the basis of the recent evolution of internet use for bank transactions: (1) there is an increase in the learning curve of banking users, who are increasingly qualified in the use of computers and "more friendly" software; (2) customers are more aware of the possibility of conducting banking transactions through this network of computers, owing to the massive advertising campaigns that banks hold in printed and audiovisual media; (3) the number of computers with internet access has significantly increased in Brazil; (4) bank systems insistently seek solutions to reduce the costs of transaction processing (Albertin 1999: 30).

The last point deserves further attention. Besides being faster and more efficient, channels with no face-to-face interaction represent a significant saving of costs for banks (especially those that use the internet). In this direction, it is interesting to compare the costs of the "informational" transactions (performed via the internet) with mechanical transactions (performed by letters and/or phone). While a letter costs as much as R$8.30 and a phone call costs R$1.84, internet transactions cost R$0.92 with direct contact to an operator, R$0.18 with the occasional intervention of an operator, and only R$0.09 as a fully automatic answer (numbers for Brazil according to Silva 2005: 84).

Melo et al.'s study (2000) indicates that there is an increased use of information systems (internet banking and home banking), and that the use of these channels have two basic purposes: (1) increase customers' accessibility to financial circuits; and (2) reduce transaction costs. According to the authors, the costs of service provisions for the banks can decrease or increase depending on the technical system used. A customer's contact in a bank branch costs US$1.07, a phone contact US$0.54, the use of an ATM US$0.27, while home banking costs US$0.15 and the internet US$0.10 per use case (ibid.: 55).

In terms of cost, there is a clear advantage in utilizing the internet and we see that the development of finance is closely linked to the development of informational techniques. The more that informational systems are used, the more the costs decrease because there are practically no workers involved. This means that these technical systems contribute to reducing the workforce in financial institutions, as they have an increasingly high degree of technical composition in its internal processes and infrastructures (Jinkings 2002).

Moreover, the effective connectivity of electronic channels has two major limitations. First, connectivity depends on the distribution of contact points of the channels (credit card terminals, personal computers, telephones, etc.). Second, connectivity of electronic channels is limited due to the regulatory differences between countries. When points of reception/transmission of banking data are spread, and when there is certain normative homogeneity between the electronic channels of countries, the fluidity of the information exchange is even stronger, meaning that the action of the bank is also more invasive and ubiquitous.

Despite this invasive potential of financial techniques, which promotes an overspread of access to financial transactions (Contel 2011), the channels still depend upon the territory contents and the physical topology of the network points. There are no actions that occur without objects, and objects only gain existence (or feature) when triggered by a particular actor (Werlen 1993). Thus, the distribution of technical systems in the given area is one of the factors that modulate the reach of access to financial services.

After these general considerations about the main geographical and technical ways to provide banking services, one may ask: How have these systems spread recently in Brazil? Which are the main forms of banking services in Brazil today? How did these technical systems affect the indebtedness of the population, especially the lower-income groups? Apparently this new technical basis, which enabled a huge reach of bank action, also allowed greater access of credit to the population—and, hence, the phenomenon of the increase of those in debt. It is important to notice that this indebtedness is particularly high among the population with the lowest income.

New channels of bank services and the spread of indebtedness

Since the mid-1980s, the Brazilian banking system has entered a new technological and organizational phase. The new Federal Constitution (1988) allowed banks to become large conglomerates (opposing the previous segmented model), which caused a decrease in the number of banks and an increase in inter-bank competition in the country. This conglomeration, along with macro-economic plans to control inflation, led to significant technological modernization in national banks. Since then, a new banking topology has evolved that is based on informational channels with no face-to-face interaction and with strong auto-mation of branches all connected to each other. As described by Diniz et al. (2010: 226), this modernization has meant that, "the user is no longer a client of an agency, but became a customer of the whole bank." The 1990s saw the defini-tive introduction of technical equipment and systems such as ATMs, magnetic cards, automatic bill payment, electronic billing, automatic investments and withdraws, IVR (Interactive Voice Response), electronic funds transfers, "cash dispensers," tele-shopping terminals, and POS systems—all of which were pro-vided in business establishments (ibid.: 227).

In addition to this new banking topology, other economic and political factors are also responsible for the recent modernization of bank topology and the indebtedness of low-income people in Brazil. Since 2003, federal policies for income distribution have been implemented and "financial inclusion" has been highlighted by the Workers Party (Partido dos Trabalhadores) with Luis Inacio Lula da Silva (2003–2011) and Dilma Rousseff (2011–2016). These policies have significantly increased the number of individuals who now have enough income to open bank accounts and demand services and products from these institutions. These distribution policies are operated via electromagnetic cards and simplified accounts. In addition to the impact of such policies, the increase in the minimum wage has led many citizens into the domain of the banking system. The minimum wage that used to be R$300.00 per month in 2005 was augmented to R$788.00 in 2015; this shows an increase of 162 percent in ten years (FEBRABAN 2015). Table 9.1 illustrates how income was distributed in recent years in Brazilian society and how it expanded to basic banking clients.

Two pieces of information in Table 9.1 help us to understand the increasing demand for banking services in Brazil. First, there was a significant increase in an intermediate class (Class C, which receives three to five minimum wages per month). This class used to total about 65.9 million people and it increased to 115.2 million (an increase of 77 percent). Second, there was a decrease in the share of the population closer to the poverty range (Classes D and E), which was 96.2 million people in 2003 and became 53.8 million in 2014.

Table 9.1 Recent income evolution of social classes in Brazilian population (in millions) 2003–2014

	2003	*2009*	*2014*
Class A	6.3	9.6	12.6
Class B	7.0	10.4	14.5
Class C	65.9	100.3	115.2
Class D/E	96.2	69.6	53.8
Total	175.4	189.9	196.1

Source: FEBRABAN 2015.[2]

Change in income distribution in the country is one of the factors that explain the diffusion of new service points and communication channels by Brazilian banks. This population started to join the bank system, consequently increasing the number of transactions. This larger demand led banks to increasingly adopt automated technical systems to deal with these new massive operations. Table 9.2 shows how the main channels for the provision of banking and financial transactions recently evolved in Brazil.

Besides such growth in locations providing banking services, it is also important to analyze the use of these systems through the number of transactions held in each one, as shown in Table 9.3.

Table 9.3 identifies recent banking developments divided into two main types: those that require the physical presence of the customer and those that allow the customer to perform remote (non-presence) operations. As indicated, there is a significant growth in the number of transactions in branches, ATMs, and banking correspondents. These three modes, which here are called "presential," increased in total by 5.8 percent. The presential modality is the fastest growing among bank correspondents, although this channel is not responsible for the largest amount of face-to-face-type transactions.

Table 9.2 Numbers of agencies, bank correspondents, ATMs, and POS in Brazil 2010–2014

	2010	2011	2012	2013	2014	% growth
Bank branches	20,049	21,405	22,290	23,046	24,866	24.0
Correspondents	195,438	218,855	239,150	244,699	207,192	6.0
ATMs	174,920	173,864	175,139	179,413	180,938	3.4
POS	3,419,377	3,515,646	4,096,428	4,451,824	4,985,804	45.8
Total	**3,809,784**	**3,929,770**	**8,462,777**	**4,898,982**	**5,398,800**	**41.7**

Source: Banco Central do Brasil BCB 2015.

Table 9.3 Recent evolution of the number of transactions per type of channel in Brazil 2010–2014 in millions

	2010	2011	2012	2013	2014	% annual growth
Presential	**18,936**	**21,154**	**22,581**	**22,994**	**23,771**	**5.8**
Agencies	7486	8718	9063	8682	8814	4.2
ATMs	8545	9262	10,102	10,627	10,953	6.4
Correspondents	2905	3174	3417	3686	3944	7.9
Remote	**12,216**	**14,389**	**17,961**	**21,576**	**25,980**	**20.8**
Internet, home and office banking	10,593	12,830	15,559	17,740	19,466	16.4
Call centers	1562	1362	1581	1546	1370	−3.2
Smartphones and tablets	61	196	821	2290	5143	202.4
Total	**31,152**	**35,543**	**40,542**	**44,570**	**46,691**	**12.4**

Source: Banco Central do Brasil BCB 2015.

The forms of service that grew the most, however, are in the remote operations category (typical of our so-called "information society"). As shown by the data, these forms increased by 20.8 percent over the last five years. A negative trend is noticed in the use of call centers, which showed a 3.2 percent decrease in transactions. The most important data of the table is the exponential growth in the use of smartphones and tablets for performing transactions: a total of 61 million transactions in 2010 increased to 5.143 million in 2014 (an increase of 202.4 percent in the period). The transactions carried out by smartphones and tablets surpassed the number of those made by call centers in 2013.

This increased importance of "remote" forms is also analyzed by FEBRABAN, which publishes regular reports on the implementation of information technology in the Brazilian banking system. According to the association, contemporary channels providing services are the main form of bank achievement, especially internet banking, mobile banking, and terminals for credit and debit cards.

This growth of non-presence methods, however, does not have as a counterpart the decline of the importance of agencies and correspondent banks in Brazilian banking topology. For FEBRABAN (2014: 24),

> the agencies have always been one of the main points of relationship between banks and customers. Because they are one of the few channels through which customers actually have personal contact with the bank, they represented a captive space for these people.

As the study shows, "the channels that allow personal contact—such as agencies and correspondents—are favored by customers for transactions of this nature, especially those with more complex processes and customized according to the customer's profile" (FEBRABAN 2014: 26).

As a result of this increased use of technology, we can understand the spread of various forms of credit in Brazil, as well as the indebtedness of the population. The first data in this regard concerns the extent that the credit assumes in relation to Gross Domestic Product (GDP). In 2002, loans represented 22 percent of the GDP. In 2007, this increased to 34.7 percent of GDP; while in 2010 this figure was 44.1 percent and, finally in the year 2014, loans represented 54.7 percent of the GDP (BCB 2015). Recent developments in the absolute volume of this financial variable also show significant numbers. As illustrated in Table 9.4,

Table 9.4 Recent evolution of the volume of credit in Brazil in R$ billions

	2010	*2011*	*2012*	*2013*	*2014*	*% annual growth*
Companies	936.9	1113.7	1294.2	1465.5	1605.4	14
Individuals	775.8	920.3	1074.1	1245.8	1412.1	16
Total	**1712.7**	**2034.0**	**2368.3**	**2711.4**	**3017.5**	**15**

Source: Banco Central do Brasil BCB 2015.

the amount of credit granted in the Brazilian economy jumped from about R$1.7 trillion in 2010 to approximately R$3.0 trillion in 2014, a significant growth of 76.2 percent in a period of five years (15 percent per year on average).

The granting of credit in Brazil grew at higher rates than most other economic variables (GDP, industrial production, employment, etc.). Credit granted to individuals has a moderately higher growth than credit given to companies—a trend that started in 1994 with the achievement of the Real Plan (Contel 2011). Such considerations allow us to say that there was a "creditization" of Brazilian territory (Santos 1993), and that this creditization is now a common phenomenon among Brazilian citizens and not only a matter for public and private companies as was seen previously.

The spread of credit and indebtedness of individuals and families is actually a worldwide phenomenon, as shown by Robert Guttmann and Dominique Plihon (2008). For the authors, this imbalance between the limits of income and consumption—that allows credit—is one of the main factors that explains the growth of the current finance-led capitalism. As noted by Ribeiro and Lara (2016: 342), "The concession of credit which was previously meant only to businesses or consumers who had proof of income or property, is now also offered for medium salaried fractions, with low income verification."

What draws our attention in this spread of debt is precisely the popularization of credit, which has been even stronger among the population with the lowest income. The main Central Bank reports on financial inclusion (BCB 2009, 2015), as well as the statistics of entities for credit risk management, have confirmed this process. The data provided by the company Serasa-Experian (2015) reveals that, in proportional terms, individuals who earn less than one minimum wage per month are those who have become most in debt. The debt index shows that people who earned less than R$500.00 per month had an indebtedness ratio of 80.7 percent in 2007, and seven years later this index jumped to 178.7 percent. However, among the population with a monthly income greater than R$10,000.00, we see an 89.3 percent debt ratio in 2007, rising to about 120.5 percent in 2014.

The official Central Bank figures (BCB 2015) confirm and explain this trend. As shown in Table 9.5, among the 56.018 million Brazilians who had some debt in 2014, 34.419 million—61.4 percent of the total—receive monthly less than three minimum wages. Although the average value of loans is not very high in proportional terms (R$11,297.00), the total balance of loans in this low-income population represents about 28 percent of the debt of individuals in Brazil today.

One last highlight of this Central Bank report is the fact that the commitment of the income of poor individuals is also greater (compared with the segments that have higher monthly income). The population earning up to three minimum wages has 24.1 percent of its monthly income committed to pay their loans. Those who earn three to five salaries have committed 23.7 percent of their income on this; those who receive five to ten salaries, spend on average 20.9 percent; and finally, those who earn more than ten minimum wages per month spend only 15.7 percent of their income paying these loans. This is a fact that

Table 9.5 Number of loans' borrowers and average value of the loans per social class in Brazil in 2014

Income	Number of clients	Average value (R$)	Percentages of total loan values
<3 minimum wages	34,419,000	11,297	28
3 to 5 minimum wages	8,202,000	24,370	14
5 to 10 minimum wages	6,807,000	40,586	20
>10 minimum wages	4,794,000	111,532	38
No income	1,796,000	–	1
Total	**56,018,000**	**25,208**	**100**

Source: Banco Central do Brasil BCB 2015.

also shows the greater vulnerability of the lower-income population in relation to the consumption of financial products in Brazil (BCB 2015).

Moreover, according to the report, the conditions imposed by the financial market to the population with the lowest income are tougher. Those without fixed income or with low income, who do not have property or credit history, tend to pay higher interest rates; this is theoretically due to the higher risks involved. The insolvency in this social group thus turns out to be a "natural" result of the high debt, or of the over-commitment of income, or a result of adverse events. When the weight of the financial commitments affects the entire budget of the citizen, or any family budget, the debt becomes a serious problem. This then leads to difficulties in covering basic fixed expenses such as rent, education, water, energy, transportation, food and bank services (BCB 2015: 125).

To understand this credit spread, it is crucial to consider the role of correspondent banking as a new channel dependent on information technology but one that also offers a face-to-face service. As shown in the 2009 Central Bank report (BCB 2009), about 85 percent of the low-income population served by federal social programs use bank correspondents to receive their benefits (about 40 million people in 2009). The bank that most uses its correspondents to make these social programs payments is the Caixa Econômica Federal (Brazil's largest public savings bank), which also offers this network for all guaranteed social benefits such as pensions, unemployment insurance, compensation for job dismissal, among others (Canton 2010: 32).

Final remarks

In seeking to identify the importance of information technology to explain the higher indebtedness of the low-income population in Brazil, the relevance of the modernization of the Brazilian macro-technic systems and the provision of banking networks became clear. This incorporation of new computerized technical systems—ATMs, banking remote networks—can be seen as one more element for dissemination of informational scientific-technical milieu (Santos 1994, 1996) in the national space.

Added to these new technologies, the distribution of income since 2003 increased the share of bank users among the population, allowing commercial banks to become ubiquitous in Brazil. This greater capillarity led to the insertion of a huge amount of people into the world of finance—either through the opening of bank accounts, the use of magnetic cards to receive benefits, the consumption of credit, or face-to-face operations in agencies. Such mechanisms increase the dependence of the population in relation to financial products and insert new contingents of people in this financial rationality that is increasingly globalized (Santos 1996). In less urbanized areas, this low-income population that becomes part of the financial system breaks with the local "pre-modern" rationality in which they previously lived. The daily access to banking and new goods results in new forms of sociability that are sometimes disembedded, transforming "everyday regionalization" of these populations (Werlen 1993).

Finally, it is worth noting that this increased capillarity of the banking system has not yet exhausted the diffusion of financial rationality in Brazil and the daily life of its population. The three main indicators that allow us to assert this are: (1) the credit/GDP ratio in Brazil is considered low compared to developed countries (where the ratio goes to 100 percent in many cases); (2) a significant part of the population does not even have a bank account so far (according to FEBRABAN only 60 percent of the population was registered with a bank in 2014); and (3) the stronger use of technology by banks will allow more customization of financial products, increasing their consumption in all income groups.

The question that remains is whether this new geography of "financial inclusion" that occurred recently in Brazil may encourage economic development and financial empowerment of individuals or if it will diffuse an even more exclusive rationality, more related to a "financial exclusion" (Leyshon 1995). If it increases this exclusion, it can increase the dependence of citizens in relation to banks, leading individuals to an even more alienated form of integration into the society of consumption, consequently preventing full citizenship to actually develop in Brazil.

Notes

1 The study of Jackeline Silva (2005: 82) listed no less than 17 electronic channels that emerged in the 1980s, which already at that time expressively increased the possibilities of banking transactions. Some of these channels, however, did not have a significant diffusion, mainly: (1) email banking; (2) financial planner; (3) fax; (4) pager; (5) interactive TV; (6) web TV; (7) video game; (8) personal digital assistant (PDA); and (9) screen phone. These systems were used only residually due to their low levels of safety and practicality, as well as owing to the high costs involved.

2 According to the Brazilian Institute of Geography and Statistics (IBGE), the income classes in Brazil may be divided as follows: Class A: monthly income of more than 15 minimum wages; Class B: monthly income from 5 to 15 minimum wages; Class C: monthly income three to five times the minimum wage; Class D: monthly income of one to three minimum wages; and Class E: monthly income of up to one minimum wage.

References

Albertin, A. L. (1999) "Comércio eletrônico: um estudo do setor bancário," *Revista de Administração Contemporânea*, 3 (1): 47–70.

Banco Central do Brasil BCB (2009) *Perspectivas e desafios para inclusão financeira no Brasil: visão de diferentes atores*, Brasília: Banco Central do Brasil.

Banco Central do Brasil BCB (2015) *Relatório de inclusão financeira (no. 3)*, Brasília: Banco Central do Brasil.

Bukharin, N. ([1915] 1969) *O imperialismo e a economia mundial*, São Paulo: Gráfica Editora Laemmert S.A.

Canton, A. M. (2010) *A Rede lotérica no Brasil*, Brasília: Instituto de Pesquisa Econômica Aplicada/IPEA.

Contel, F. B. (2011) *Território e finanças: técnicas, normas e topologias bancárias no Brasil*, São Paulo: Annablume.

Dias, L. (1995) *Réseaux d'information et réseau urbain au Brésil*, Paris: L'Harmattan.

Diniz, E., Correia da Fonseca, C. E. and de Souza Meirelles, F. (2010) *Tecnologia bancária no Brasil: uma história de conquistas, uma visão de futuro*, São Paulo: FGV.

FEBRABAN (2014) *Pesquisa Febraban de tecnologia bancária*, São Paulo: FEBRABAN.

FEBRABAN (2015) *Painel econômico financeiro*, São Paulo: FEBRABAN.

Giddens, A. (1991) *As consequências da modernidade*, São Paulo: Editora Unesp.

Guttmann, R. and D. Plihon (2008) "O endividamento do consumidor no cerne do capitalismo conduzido pelas finanças," *Economia e Sociedade*, 17: 575–610. http://dx.doi.org/10.1590/S0104-06182008000400004.

Jinkings, N. (2002) *Trabalho e resistência na "Fonte Misteriosa": os bancários no mundo da eletrônica e do dinheiro*, Campinas: Editora da Unicamp/Imprensa Oficial do Estado.

Labasse, J. (1976) *L'espace financier: analyse géographique*, Paris: PUF.

Leyshon, A. (1995) "Geographies of money and finance I," *Progress In Human Geography*, 19 (4): 531–543.

Melo, P. R., Carlos, E., Rios, S. D., and Gutierrez, R. M. (2000) "Os mercados de automação bancária e comercial," *BNDES Setorial*, 11: 47–70.

Ribeiro, R. and Lara, R. (2016) "O endividamento da classe trabalhadora no Brasil e o capitalismo manipulatório," *Serviço Social & Sociedade*, 126: 340–359.

Santos, M. (1993) *A urbanização brasileira*, São Paulo: Hucitec.

Santos, M. (1994) *Técnica, espaço e tempo: globalização e meio técnico-científico informacional*, São Paulo: Hucitec.

Santos, M. (1996) *A natureza do espaço: técnica e tempo, razão e emoção*, São Paulo: Hucitec.

Santos, M. and Silveira, M. L. (2000) *Brasil: território e sociedade no início do século XXI*, Rio de Janeiro: Record.

Serasa-Experian (2015) "Indicador serasa experian de demanda do consumidor por crédito," available online at: http://noticias.serasaexperian.com.br/indicadores-economicos/demanda-do-consumidor-por-credito/ (accessed April 28, 2016).

Silva, C. A. (1999) "As transformações da rede de gestão territorial do Banco Nacional S/A sob a égide da revolução telemática," *Território*, 6: 55–72.

Silva, J. (2005) *Sistemas de redes de transações bancárias: uma análise sob as óticas tecnológica e jurídica*, São Paulo: Escola Politécnica/USP (Dissertação de Mestrado).

Werlen, B. (1993) *Society, action and space: an alternative human geography*, London: Routledge.

10 Digital health mapping

Big data utilization and user involvement in public health surveillance

Annika Richterich

Introduction

In the aftermath of the 2010 Haiti earthquake, a group of volunteers collaboratively utilized the Ushahidi mapping platform in order to create a cartographic overview of the disaster's devastating ramifications: they mapped information such as citizens' medical requirements, water availability and needs, as well as public health reports. Ushahidi, which means "testimony" in Swahili, is an open source project that originated in Kenya and enables users to create crowdsourced maps of humanitarian or ecological crises. After the earthquake in Haiti, the service was used in order to collect and geolocate information which was crucial for delivering and maintaining effective humanitarian aid programs (Morrow et al. 2011). Relevant, timely data were derived from sources such as the micro-blogging platform Twitter and a free text messaging service which citizens were offered in order to communicate their most urgent needs and location. The emerging public map supported aid workers in assessing where help was needed and what kinds of (medical) supplies were required. It also provided those affected—as well as their relatives and friends worldwide—with desired updates. More than two years later, Patrick Meier, one of the initiators of the Ushahidi-Haiti project, commented on the platform's utilization as well as the project's reception. While he briefly mentioned the positive feedback that the project received, his blog post also indicates his frustration with its recurring criticism: He accuses critics of the project of being "masters at smart-talk" (Meier 2014) rather than "offering constructive criticism."

In particular, this comment refers to the dilemma that sensitive, personal information was collected and mapped during the project (in a situation of extreme urgency), while explicit consent could rarely be asked. Haitian citizens were in a position in which privacy concerns were outweighed by existential needs, hence gladly turning to offered services. In his post, Meier presents an activist position that calls for constructive engagement and a "proactive" approach to digital humanitarian technology. One is left wondering how—in light of humanitarian benefits of a project, for instance—researchers shall raise and communicate criticism and possibly facilitate an improvement of digital activist approaches? This question is also relevant to this chapter, since it will discuss emerging, digital methods in public health mapping. In particular, it aims at providing a constructive assessment of health mapping projects involving multiple "big data" sources as well as volunteer contributions.

Digital tools for public health (crises) mapping—e.g., regarding the outbreak of cholera after the 2010 Haiti earthquake—have received increased attention during the last years (see, e.g., Meier 2012a; Soden and Palen 2014; Ziemke 2012;). Social networking sites, blogging platforms, or news websites have shown to be insightful sources for retrieving data indicating citizens' health status (Chunara et al. 2012). Monitoring and harnessing such sources can hence facilitate targeted aid and interventions. These approaches are not merely initiated and maintained in moments of health crisis. Rather, in addition to projects founded in reaction to humanitarian crises, one could also witness the emergence of long-term efforts in mapping public health developments. Particularly in the field of public health surveillance, dedicated to monitoring the spreading of infectious diseases, new approaches in collecting and mapping health data have emerged. Biomedical disciplines such as epidemiology and its branch public health surveillance have conventionally relied on data from clinical and virological diagnosis or mortality rate statistics. However, recent technological developments and emerging big data approaches have created new types of digital health data: The information may be derived from social networking sites, news wires, sensors, or web search logs.

The emerging possibilities for big data retrieval and analysis have enhanced the field of public health surveillance and facilitated new forms of online services concerning public health. Among the projects relevant to this field are the websites Healthmap.org and FluNearYou.org. HealthMap was launched in 2006 and has been developed by a team of researchers, epidemiologists, and software developers at the Boston Children's Hospital (U.S.). The website monitors several digital sources such as social media, news websites, or professional reports issued for instance by the World Health Organisation (WHO). It provides an overview of selected content in a public (Google) map (Brownstein et al. 2008; see Figure 10.1). FluNearYou is closely related to the HealthMap project; it was developed by partly the same researchers. The project was originally created by epidemiologists at Harvard, the Boston Children's Hospital, and The Skoll Global Threats Fund. Its creators have described the service as an example for "participatory epidemiology" (Freifeld et al. 2010; see also Jost et al. 2007). It is based on symptom reports submitted by North-American users, whose submissions are subsequently located on a publicly available online map (see Figure 10.2). Both projects are available as websites and as smartphone apps. The two platforms have been selected as case studies to be analyzed in this chapter, since they are ongoing examples of public health mapping services (July 2016) that were extensively documented by involved researchers.

Innovative approaches in digital public health surveillance also come along with new challenges for source selection, data retrieval, and visualization. With regards to the collection of data in "participatory projects," questions concerning an encouragement of user involvement as well as the verification of information

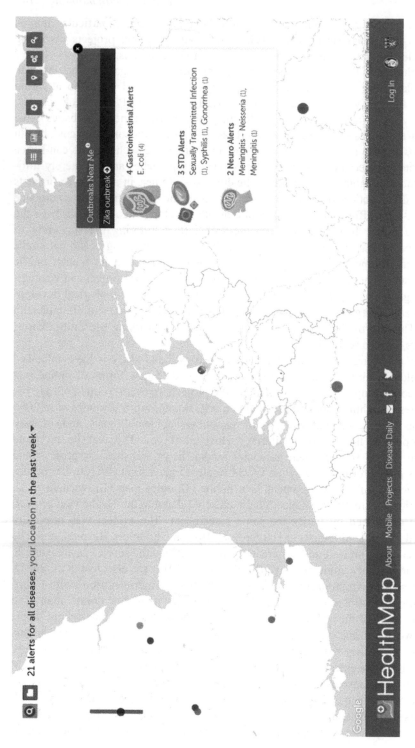

Figure 10.1 HealthMap disease alerts for the location "Amsterdam, Netherlands."

Source: Screenshot taken by the author on July 12, 2016 from www.healthmap.org/en.

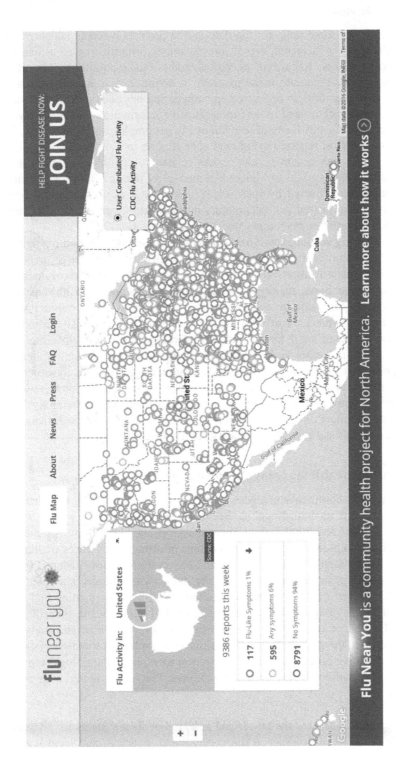

Figure 10.2 The "Flu Map" of the North American community health project FluNearYou.

Source: Screenshot taken by the author on July 12, 2016 from https://flunearyou.org.

are crucial. At the same time, for big data-driven projects it seems insightful to question the nature of selected, digital sources and algorithmic as well as human criteria for this choice. Regarding the visualization of data, public services often utilize geographic, corporate maps in order to communicate potential disease intensities and spreading. In order to map content that has been classified as relevant, these approaches need to determine locative information for the retrieved data. Certain reports may be mapped without any geographic uncertainty, e.g., in cases where a person reported symptoms with regards to a distinctive location. However, other instances may raise issues related to spatial uncertainty and ambiguity. In order to assess to what extent innovative public health surveillance services have been affected by such issues, this chapter examines: which data sources have been utilized in contemporary digital health mapping projects, and how has this information been translated into spatial relations?

In the following section, I will first contextualize the phenomenon of digital, public health mapping with particular regards to neogeographic user involvement, and big data utilization. Subsequently, I will present an analysis of mapped health information based on the aforementioned case studies HealthMap and FluNearYou. By analyzing the interfaces and excerpts of available online maps, I will investigate what kind of information has been geographically located in what ways. In doing so, I aim at (at least partially) reverse-engineering the decision-making processes and algorithmic criteria which underlie these health mapping projects. Hence, the analysis sheds light on the geographic construction of public health developments in digital mapping projects and points toward challenges concerning data selection, retrieval, assessment, and authenticity. My conclusion will reflect on the insights presented in my analysis and will likewise point out some of the directions in which further research is needed.

Digital public health mapping: neogeography meets big data

The year 1854 is commonly named when it comes to determining the birth of epidemiology and its branch public health surveillance (see Newsom 2006; Paneth 2004; McLeod 2000).[1] Back then, the physician John Snow set out to map the cholera epidemic in London, locating information on cases of illness and death. When mapping his diagnoses and observations, Snow deduced from the increased frequency of infections close to public water pumps that contaminated water caused the disease and its spreading. Still today, spatial analyses and visualizations of epidemics are part and parcel of public health surveillance (see also Ostfeld et al. 2005). Already in the mid-1990s, Clarke and Langley (1996) examined the potential use of emerging geographic information systems (GIS), i.e., locative processing and visualization tools, in epidemiology. On the one hand, the authors highlighted the promises that come along with such developments:

GIS applications show the power and potential of such systems for addressing important health issues at the international, national, and local levels.

Much of that power stems from the systems' spatial analysis capabilities, which allow users to examine and display health data in new and highly effective ways.

(Clarke and Langley 1996: 85)

On the other hand, while seeing GIS as potentially powerful tools, in their conclusion the authors also reflect on the challenges posed by such new technologies:

While it holds distinct promise as a tool in the fight against emerging infectious diseases and other public health problems; it is not simply the next widget to come into play. GIS can be seen as a new approach to science, one with a history and heritage, a finite and well researched suite of methods and techniques, and a research agenda of its own. It does not fit neatly into the health scientist's toolbox. It requires rethinking and reorganizing the way that data are collected, used, and displayed.

(Clarke and Langley 1996: 91)

In their early assessment of GIS use for digitally supported public health surveillance, the authors hence already address issues related to adjusting newly available technologies to the needs of this biomedical discipline. While this was initially mainly related to new analytic possibilities and visualization techniques such as layers and multimedia elements (ibid.), these reflections were likewise applicable to later technological advancements. With regards to this chapter, two developments are particularly relevant: first, the possibility to involve users in data collection and mapping efforts by utilizing online systems that allow for user-generated content (see O'Reilly 2007; for critical perspectives see Petersen 2008; Van Dijk 2009); second, digital technologies which have enabled new forms of data collection, commonly described with the term and (former) buzzword "big data." Also in the field of epidemiology, such sources have been explored as basis for digital disease detection (Brownstein et al. 2009). Both aspects will be explored in the following sections, since they present seminal developments that were also relevant to selected case studies.

Neogeography and participatory epidemiology

While John Snow had to collect and map relevant health data himself, contemporary approaches allow for an active involvement of users. Digital technologies have facilitated a trend toward "neogeography," i.e., the involvement of amateur-cartographers in map-making practices. This development was likewise relevant to the field of epidemiology. As Sanjay and Joliveau state, while the definition of neogeography may be somewhat contested among its proponents,

most researchers would agree on the common keywords that have come to be associated with the works on NeoGeography. These keywords include

Web 2.0, mashups, public participation, social networking, volunteered geographical information, crowd-sourced data, user-generated content, open
source maps API [Application Programming Interface] and affordable navigational devices, principally GPS data loggers.

(Sanjay and Joliveau 2009: 75)

More generally, the authors describe neogeography as "another outcome of the
increasingly close integration of our lives with geocomputational and World
Wide Web technology" (Sanjay and Joliveau 2009: 79; see also Goodchild
2009; Turner 2006). In this sense, neogeography needs to be placed in the
context of wider debates concerning an allegedly more "participatory culture"
enabled by Web 2.0 platforms and allowing for direct user engagement and
many-to-many communication (see, e.g., Burgess and Green 2009; Deuze
2006; Jenkins 2006). There are various mapping services—such as Google
(My) Maps, the initially mentioned Ushahidi platform, or the free and open
source project OpenStreetMap—which enable users to map information or
even create the cartographic surfaces as such. It has been emphasized though
that these participatory mapping approaches are still subject to regulations
defined by the map hosts. This is particularly relevant in cases where the cartographic material is owned by corporations such as Google Inc. In consequence, various authors have advised caution when it comes to asserting a
"participatory mapping culture" and its democratization, since neogeographic
practices are for instance defined by access to the internet and digital content
as well as digital skills and literacy. Haklay (2013: 55) stresses with regards to
the promise that neogeography "is for anyone, anywhere, and anytime" that
looking at the actual practices exposes that "there is a separation between a
technological elite and a wider group of uninformed, labouring participants
who are not empowered through the use of the technology."[2] In addition to
such issues of accessibility and expertise, there are new forms of dependencies
which are related to the dominance of global media corporations: "One of the
more curious aspects of NeoGeography is the high dependency of much activity on the unknown business plans of certain commercial bodies providing
API's for mapping" (Sanjay and Joliveau 2009: 80). This also has an influence
on the sustainability of projects relying on commercial APIs, since the conditions for using them may change.

This trend toward neogeography and related concerns are relevant to public
health surveillance, since researchers and practitioners in this field are exploring
possibilities to involve amateur users in monitoring and mapping infectious
disease developments. With computer technology and the Web 2.0 coming of
age, we can identify different strategies related to the types of technologies
utilized for user engagement: On the one hand, there are websites such as the
European InfluenzaNet (www.influenzanet.eu), FluNearYou, or the Grippeweb
(German for Flu-Web; see https://grippeweb.rki.de) which require a community
of volunteers with internet access, their willingness, and the capacity to report
observed symptoms. On the other hand, the increased popularity of smartphones

and tablets has made it attractive to utilize apps for an involvement of users. As Meier (2012a: 89) puts it with regards to projects dedicated to crisis mapping: "Indeed, human beings equipped with mobile phones make for formidable mass multimedia sensors." Research and projects drawing on mobile technologies for health-related objectives have been described as mHealth approaches (Kay et al. 2011; Lupton 2013). From a rather critical perspective, Lupton (2013: 393) observes that

> [m]any articles have recently appeared in the health promotional and pre-ventive medicine literature, ruminating on the possibilities of being able to communicate with the public, monitor their behaviours and conduct health promotion interventions via the mobile devices that they carry with them or wear throughout their day.

Regarding the field of public health surveillance and possibilities of "participatory epidemiology," Freifeld et al. (2010) present an assessment of their smartphone/tablet app OutbreaksNearMe. The researchers developed this app as element of HealthMap and describe the technology as a step toward future systems "engaging the public as participants in the public health process."[3] The authors also identified pioneers and examples of public health surveillance aimed at involving volunteers, many of which are based on users sending SMS with relevant health information. Among them are the FrontlineSMS platform,[4] the aforementioned Ushahidi mapping of users' emergency text messages, and the GeoChat collaboration software which needs to be installed on Mac or PC, but also involves users' SMS input.

Big data and digital health

Neogeographic efforts in public health surveillance are accompanied by attempts to diversify and amplify data used for monitoring infectious disease developments. Services such as HealthMap do not merely present the input of users but are likewise retrieving data from other various online sources. Therefore, endeavors to digitally map and monitor infectious disease development are likewise closely related to the field of data science. The continuous documentation and archiving of digital user behavior, especially by large corporations, and the immense amounts of data which are being produced in this process are commonly labeled "big data": a term which has been as widely used as it has been criticized—e.g., for being a meaningless buzzword (see, e.g., Rao 2013) and for fostering a "digital positivism" (Mosco 2015).

Traditionally, data have been scarce and their compilation in research was subjected to controlled collection and deliberate analytical processes (Boyd and Crawford 2010; Kitchin 2014a). In contrast, the

> challenge of analysing Big Data is coping with abundance, exhaustivity and variety, timeliness and dynamism, messiness and uncertainty, high

relationality, and the fact that much of what is generated has no specific question in mind or is a by-product of another activity.

(Kitchin 2014a: 2)

Big data differ from traditional large-scale datasets with regards to their volume, velocity, and variety (Boyd and Crawford 2012; Kitchin 2014a, 2014b; Marz and Warren 2012; Zikopoulos et al. 2012). Moreover, such datasets are comparatively flexible, easily scalable, and have a strong indexical quality. While volume, velocity, and variety are commonly used to define big data, critical data scholars such as Lupton (2015: 1) have highlighted that "[t]hese characterisations principally come from the worlds of data science and data analytics. From the perspective of critical data researchers, there are different ways in which big data can be described and conceptualised." In her outline of a critical sociology of big data (Lupton 2014), the author describes big data as knowledge systems that are embedded in and constitute power relations. Echoing Lupton's criticism of dominant approaches to understanding big data, this chapter is likewise meant as a contribution to critical data studies (see also Dalton and Thatcher 2014; Dalton et al. 2016).

Approaches in the field of big data-driven public health surveillance can be broadly categorized according to the kind of data used and the ways how these have been retrieved. Especially in its earlier days, digital disease detection particularly focused on publicly available online sources and monitoring: For example, online news websites were scanned with regards to information relevant to public health developments (Eysenbach 2009; Zhang et al. 2009). With the popularization of social media, it seemed that epidemiologists no longer had to wait until news media were publishing information about potential outbreaks. Instead, they could harness digital data generated by decentralized submissions from millions of social media users worldwide (Eke 2011; Velasco et al. 2014). Especially platforms such as Twitter which allow for access to (most) users' tweets through an open application programming interface have been explored as useful indicators of digital disease developments (Signorini 2011; Stoové and Pedrana 2014). In particular, there were and are several attempts at combining social media and news media as sources (Chunara et al. 2012; Hay et al. 2013), as well as projects which utilized search engine queries in order to monitor and potentially even predict infectious disease developments. For example, the platforms EpiSPIDER[5] (Keller et al. 2009; Tolentino et al. 2007) and BioCaster (Colier et al. 2008) combined data retrieved from various online sources, such as the European Media Monitor Alerts, Twitter, reports from the U.S. Centers for Disease Control and Prevention and WHO. The selected information was then presented in Google Maps mashups. Meanwhile however, these pioneer projects seem to have been discontinued, while the HealthMap platform—which will be discussed in more detail below—is still active (see Lyon et al. 2012 for a comparison of the three systems).

Likewise, big data produced by queries entered into search engines were utilized for public health surveillance projects. In particular, studies by Eysenbach (2009),

Polgreen et al. (2008) and Ginsberg et al. (2008) have explored this aspect of *infodemiology*. The authors demonstrated that Google and Yahoo search engine queries may indicate public health developments, while they likewise point to methodological uncertainties caused, for example, by changes in users' search behavior. Moreover, such approaches using search engine data have been described as problematic, since they are based on very selective institutional conditions for data access and have raised questions concerning users' privacy and consent (Lupton 2014: 93ff.; Richterich 2016). In this context it is also indicative that a project such as Google Flu Trends—which was initially perceived as "poster child of big data"—was discontinued after repeated criticism (Lazer and Kennedy 2015; Lazer et al. 2014). In light of such developments and public concerns regarding big data utilization (Science and Technology Committee 2015; Tene and Polonetsky 2012: 251ff.), ethical considerations are receiving more attention (Mittelstadt and Floridi 2016; Vayena et al. 2015).

While having been meanwhile discontinued, Google Flu Trends is an illustrative example that highlights how collaborations between epidemiologists and data and computer scientists have facilitated applied research possibilities. Similarly, some of the aforementioned authors—such as Brownstein, Freifeld, and Chunara—have also been involved in projects that used their insights in order to develop new applications. Among them are the websites and mobile applications of the HealthMap project and FluNearYou. It is characteristic for these services that health information is presented in forms of maps and based on neogeographic methods as well as big data retrieval. More specifically, HealthMap draws on multiple data sources, combining big data with user-generated content, while FluNearYou is an example of "participatory epidemiology" and presents submissions from registered community members.

Case studies

HealthMap and FluNearYou will be examined in more detail in the following section. I will particularly focus on two questions: First, what kind of information is being mapped? Second, how is information mapped, i.e., according to which criteria are geographic locations chosen and hence indicated on the map? Both platforms are examples for projects in which epidemiological research, data science, and bioinformatics go hand-in-hand. They have been developed by interdisciplinary teams of epidemiologists, computer scientists (particularly bioinformaticians), and data scientists. The platforms are based on research that investigates how effectively certain online sources may be used in order to monitor and predict infectious disease developments. Subsequently, these investigations have been put into practice by creating websites and mobile applications presenting results and examined methods to the public.

The two projects are closely related: HealthMap was launched in 2006, and while an interdisciplinary team at Boston Children's Hospital is and has been involved in its development and maintenance, epidemiologist John Brownstein as well as computer scientists and biomedical engineer Clark Freifeld have leading roles in this process. The project has been extensively documented by researchers involved in its development (see, e.g., Brownstein and Freifeld 2007; Brownstein et al. 2008; Freifeld et al. 2008).[6] It receives funding from multiple corporations, such as Amazon, Google, and Twitter. Visually, the interface is dominated by a Google Map: in this map, health-relevant information—such as news items on disease outbreaks or tweets concerning disease developments in a certain region—are located. Depending on the amount of relevant reports and their assigned significance, smaller or larger dots may be clicked by the user in order to receive: first, an overview of one or multiple sources selected as relevant for a country or region; second, more detailed information on and a link to the original source. The selection process is automatized, i.e., certain sources are monitored by default and an algorithm determines which content will be included. Depending on the website user's location, a smaller text-box indicates potential "Outbreaks in current location" which are clustered into 12 disease categories. On July 20, 2016, out of 889 alerts from the past week, the main part of reports was related to "Vector-borne Alerts" (742 in total; e.g., Dengue and Zika virus), and "Respiratory Alerts" (202 in total; e.g., Influenza H1N1 and Tuberculosis). In July 2016, the website put particular emphasis on the 2016 Zika virus pandemic.[7] The virus is dominantly spread by Aedes mosquitos and causes symptoms such as (mild) fever, muscle pain, or headache. Infections are particularly harmful in case of pregnant women, since the virus can cause birth defects.

FluNearYou is an initiative of HealthMap (see HealthMap Brochure n.d.: 2). It was launched in October 2012 and has been developed in collaboration with the American Public Health Association and the Skoll Global Threats Fund (see, e.g., Chunara et al. 2013; Wójcik et al. 2014). Similarly to HealthMap, a main element of the website is its "Flu Map": a Google Maps interface that shows user reports on influenza/influenza-like-illness symptoms for the United States and Canada. The website states that it is supported by a community of more than 60,000 "flu trackers." On Wednesday, July 20, 2016, it had received 2978 reports that week so far (all users receive a weekly reminder to participate on Mondays).

The insights described in the following sections are based on observations of the platforms conducted in July 2016. Of course, one has to keep in mind that, due to climatic factors, seasonal influenza activity in the Northern Hemisphere commonly peaks between December and February (see Centers for disease control and prevention 2014). Therefore, during the time frame relevant to the analysis below, potential influenza activity as indicated by both services was expectedly low. This is particularly relevant to FluNearYou. However, one has to take into account that it is not the aim of this chapter to assess the effectiveness and performance of the observed websites; therefore, timing the observations with influenza season was not part of my approach.[8] Instead, I will mainly examine which information has been mapped and according to which criteria geographic locations have been determined. In addition to my own observations, I have also included information on the two websites provided by prior reports.

Identifying health data

In order to understand the cultural geographies of public health that are con-structed through the aforementioned projects, it is vital to first analyze what kind of information is being mapped and which sources are left out. HealthMap com-bines data which are retrieved by scanning multiple sources such as the commer-cial news feed aggregators Google News, Moreover (by VeriSign), Baidu News, and SOSO Info (the last two are Chinese language news services), institutional reports from WHO and the World Organisation for Animal Health, as well as social news media such as Twitter.[9] The platform utilizes global sources and is not limited to a particular country. However, as the following section illustrates even more clearly, it depends on a sufficient, reliable basis for retrieved informa-tion—which is usually not guaranteed to an equal extent for different countries/regions.

Dominantly, the website draws on sources that are authored by public health institutions or news outlets/journalists. Before being published, such sources are commonly subject to selection and verification processes during which the quality and correctness is assessed. This goes particularly for organizations such as WHO but is also the case for quality journalism outlets (Shapiro et al. 2013). In contrast, microblogging platforms such as Twitter also contain information from individual users that may be timelier but likewise more difficult to verify (Hermida 2012). Apart from automatically retrieved social media content, users can also send individual reports: this can either be done through the website's "Add alerts" function (which is part of the top menu), by email, text message, phone call (hotline), or by using the mobile app Outbreaks Near Me.

News items are a particularly dominant type of data that is mostly retrieved from news aggregators, with Google News items being particularly prevailing. Therefore, being included in such aggregators enhances the chance for (health-indicative) news items to be presented in Health Map. With regards to "informa-tion politics" (Jordan 2015), there is hence a tendency to monitor sources that are maintained by global media corporations. These sources play an important role as gatekeepers, defining in- and exclusion. In this sense, research concern-ing the gatekeeping function of such aggregators is highly relevant to projects such as HealthMap and may be used in order to assess to what extent such an approach is (in-)appropriate (Weaver and Bimber 2008). While drawing on news aggregators seems to be a technically feasible or preferable solution, this approach raises questions regarding the selection criteria relevant to utilized big data sources. The presented data go through multiple forms of automated selec-tion: First of all, they are defined by the Google algorithm determining more generally which sources are included in its News service. Second, they are subject to an automated process in which the HealthMap algorithm selects information that is considered as relevant for disease detection.

While the algorithm as such is not accessible to external researchers, one can get some insights into the selection criteria by looking at the indicated data scheme: Information retrieved from relevant sources is clustered according to multiple categories which are shown in the "List view" (see Figure 10.3). The following categories are indicated here: source, date, summary, disease, location, species, cases, deaths, and significance (rated from 1–5). In particular, disease names and symptoms seem to function as relevant key words for selecting sources which are then included in HealthMap. In this sense, the platform is most likely—in its current state—a potentially insightful tool for monitoring known diseases, but unknown symptoms might facilitate the risk that unexpected disease developments are overlooked. However, based on an interface analysis, this argument cannot be addressed with certainty (since one would need to know which criteria/key words are decisive for selecting sources). This point also hints at an issue that has been stressed with regards to an earlier report published on HealthMap: On March 14, 2014, the site selected and mapped a report on a "mysterious haemorrhagic fever" killing eight people in Guinea (Asokan and Asokan 2015; Kotz 2015). Only several days later, on March 23, 2104, WHO published a first official report on the Ebola outbreak. Hence, even in a case where relevant, reliable information has been selected, the challenge remains to identify these reports as significant. In this sense, a highly relevant aspect of HealthMap is not only which data are being selected, but also by whom and how the website is being monitored.

FluNearYou presents two kinds of data: user reports on influenza/influenza-like-illness symptoms and influenza intensities reported by the U.S. Centers for Disease Control and Prevention (CDC). The website's FAQ section emphasizes that mapped data are based on users' reports of symptoms; it is not possible to indicate eventually confirmed cases. CDC Flu activity is categorized according to "minimal," "moderate," and "high." User reports are clustered in three categories: "no symptoms," "any symptoms," "flu symptoms." Being explicitly labeled as "community health project for North America," only residents of the United States and Canada are eligible for participation. Due to the restriction that "You must be a resident of the U.S. or Canada" in order to sign up, I have mainly limited my analysis to the publicly available interface. However, images of the mobile user interface can be found in Smolinski et al. (2015: 2125), and according to these images, users are asked to indicate symptoms such as fever, fatigue, cough, nausea, short breath, chills, headache, or body ache as well as information on their "flu vaccination." Moreover, before signing up for the service, users are asked to state a zip code within the U.S. or Canada, their age, and sex. The zip code is then used for mapping the provided information—which brings me to the next part of my analysis.

Mapping health data

How is information mapped, or more specifically, which factors are decisive for defining which location is selected and hence indicated on the respective map? In order to address this question, I analyzed the map by starting from a zoomed-out perspective and by subsequently zooming-in on individual countries. On July 20, 2016, the website indicated "889 alerts from past week" (globally; see Figure 10.4),

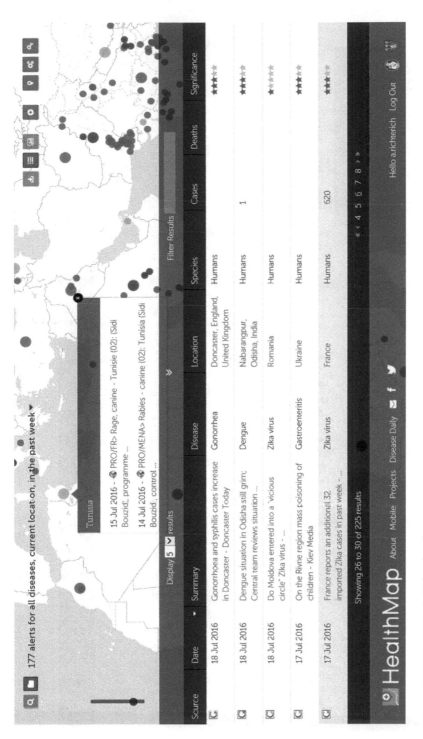

177 alerts for all diseases, current location, in the past week ▾

Tunisia

15 Jul 2016 - PRO/FR> Rage, canine - Tunisie (02): (Sidi Bouzid), programme ...

14 Jul 2016 - PRO/MENA> Rabies - canine (02): Tunisia (Sidi Bouzid), control ...

Display 5 results

Filter Results

Source	Date	Summary	Disease	Location	Species	Cases	Deaths	Significance
	18 Jul 2016	Gonorrhoea and syphilis cases increase in Doncaster - Doncaster Today	Gonorrhea	Doncaster, England, United Kingdom	Humans			★★★★☆
	18 Jul 2016	Dengue situation in Odisha still grim; Central team reviews situation ...	Dengue	Nabarangpur, Odisha, India	Humans	1		★★★☆☆
	18 Jul 2016	Do Moldova entered into a "vicious circle" Zika virus - ...	Zika virus	Romania	Humans			★☆☆☆☆
	17 Jul 2016	On the Rivne region mass poisoning of children - Kiev Media	Gastroenteritis	Ukraine	Humans			★★★★☆
	17 Jul 2016	France reports an additional 32 imported Zika cases in past week - ...	Zika virus	France	Humans	620		★★★☆☆

Showing 26 to 30 of 225 results

« ‹ 4 5 6 7 8 › »

HealthMap   About   Mobile   Projects   Disease Daily

Hello a.richterich   Log Out

Figure 10.3 HealthMap's "List view."

Source: Screenshot taken by the author on July 19, 2016 from www.healthmap.org/en.

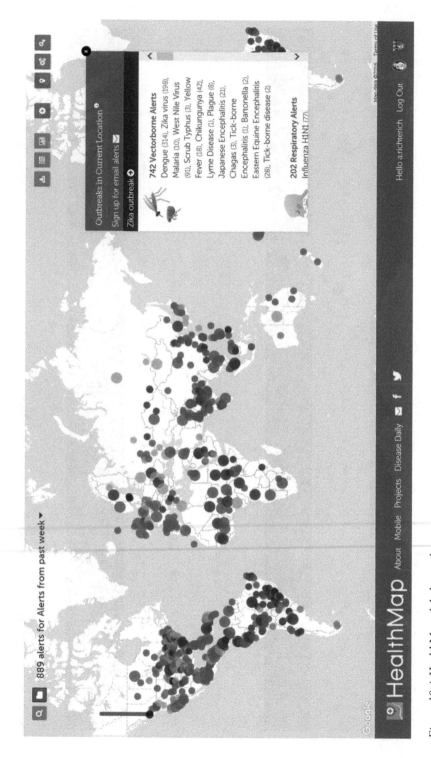

Figure 10.4 HealthMap global overview.

Source: Screenshot taken by the author on July 20, 2016 from www.healthmap.org/en.

with "Vector-borne Alerts" being dominant (742). When looking at the world map, the first striking aspect concerns the distribution of reports. While countries such as the United States (particularly Eastern states and "technophile" California) and India are comparatively densely covered, there are areas that contain fewer reports, or even none. Among the blank spaces and hence countries for which no reports have been found or submitted are Libya, Algeria, Botswana, Madagascar, Mauritania, Zimbabwe, and Zambia. The lack of information concerning African countries, for instance, is not merely a reflection of population densities, but also related to debates on a digital divide. In this context, the term refers to geographically determined inequalities regarding citizens' digital technology access more generally and internet availability in particular (Norris 2001; Sui et al. 2013; Warschauer 2004; Warf, this volume). Factors such as citizens' possibility to access the internet, a country's accessibility for aid workers and public health professionals, and a well-functioning, free press are crucial in order to ensure the effectiveness of HealthMap. In this sense, the system is only as good as the news and information basis available for individual countries.[10] The implications of this insight are at least twofold: On the one hand, the blank spaces hint at information spheres, or the lack thereof, which seem to fall through the criteria relevant to the HealthMap algorithm. It might be the case that there are simply no data sources available, but it is also possible that available data sources are not yet covered with applied selection criteria. Hence, the blank space may be used as a starting point to identify countries for which the current big data system is not effective yet. These countries could be used as targeted cases for evaluating which additional criteria could further strengthen the system and its data selection. On the other hand, one should also not completely neglect the information value inherent to these blank spaces. Apart from allowing for insights into digital disease detection, the map also acts as an indicator for needed investments in disease reporting as well as efforts aimed at tackling the digital divide.[11]

Locations selected for each report appear to be based on the textual, spatial information mentioned explicitly in a selected article; e.g., names of cities, states, or districts. When zooming-in on different countries, it is striking that certain reports are either located more broadly within that country or more specifically with regards to a distinctive district or even a city/town. For example, while reports for countries such as France or Norway were summarized under one, country-wide location mark, reports for England or Spain were geographically more specific. This observation hints at the fact that reports in some countries may be (at times) more or less likely to contain exact spatial information. In order to retrieve overall more fine-grained and information that is equally geographically specific, HealthMap depends on news reporting that includes concrete geographic information. More distinctively located information seems particularly relevant in order to enhance the platform's value as a monitoring tool for public health professionals. Likewise, geographically more specific information makes it easier for users to assess the individual relevance of reports. In addition to efforts aimed at ensuring that the algorithm picks up on sources which contain concrete geographic information and subsequently extracts those correctly, attempts at specifying the geographic significance of reports could further improve the website. This issue could be (at least partly) addressed by adjusting the visualization scheme: instead of visualizing certain reports in one location marker on the map, the scope of the report could be highlighted, for instance, in form of a more comprehensive layer. Another option could be to encourage users to specify for which areas a report may be especially relevant. However, the website already aims at involving users by allowing them to suggest reports, and this feature is rarely used.[12] Hence, such an approach would require further investment in promoting user involvement—a challenge which is at the heart of FluNearYou, a public health mapping service which has been described as example of "participatory epidemiology" by its creators (Freifeld et al. 2010).

FluNearYou geographically locates users' symptom reports: This process is based on a registration of users during which they indicate their zip code. The platform is merely open to U.S. and Canadian residents. The "Flu Map" is most densely populated within the U.S.; Canadian users were comparatively less active. On July 20, 2016, there were 3649 reports submitted for the U.S. and only 12 reports for Canada. This is most likely related to the fact that the website received extensive television news coverage in the U.S. during the 2012/2013 and the 2013/2014 influenza seasons. Within the U.S., contributions are predominantly submitted in urban areas with particularly densely mapped reports along the West coast and in Eastern states. While of course less severe than in the case of the above-mentioned issue of a digital divide, this distribution may not only reflect factors such as population densities in rural and urban areas, but also the likeliness and ease with which digital technologies are utilized by citizens. However, this cannot be assessed with certainty.

This unequal distribution—within the U.S. and in comparison to Canada—refers to one of the main challenges for "participatory epidemiology": the encouragement of user involvement and investments in community building. The platform relies on users' correct and sincere reports but also their willingness to contribute regularly. For example, it has been observed that users' willingness to contribute might be higher while experiencing symptoms (see also Smolinski et al. 2015). With regards to mapping users' contributions, it is moreover not merely relevant that they submit authentic symptom reports, but that they are willing to include the correct zip code.[13]

A crucial issue resulting from the fact that participants' contributions are located based on the zip code is that the "Flu Map" did not take into account an eventual mobility of users. This problem was likewise mentioned in a paper by researchers involved in the platform's development (Smolinski et al. 2015: 2129). It is not explained in more detail why the project focuses on the home location as indicator of potential influenza developments. One technical reason might be that this is the most feasible option; methodologically, this might also be recommendable, since participants could forget to adjust their location continuously.

Since it is focused on contributions located in U.S. territory, the platform is exclusively valuable for monitoring national disease developments. Due to the importance of efforts invested in community building, participatory approaches such as FluNearYou are less easily scalable and transferable. Nevertheless, projects such as the European InfluenzaNet and national projects such as the German GrippeWeb suggest that an exploration of this field of public health surveillance seems promising on different scales and in different national contexts. In order to support and refine such attempts at participatory epidemiology, further research is needed which investigates factors influencing users' willingness to get involved and sheds light on legal as well as ethical aspects of these innovative approaches to data collection.

Conclusion

Projects such as HealthMap and FluNearYou indicate socio-technical changes in the field of public health surveillance. On the one hand, technologically, they demonstrate that data sources in the field of epidemiological surveillance are changing: Innovative approaches in this field explore the possibility to develop and utilize algorithms for an automated retrieval of health-indicative big data from multiple, digital sources. Among these sources are for instance news items on contemporary disease developments or reports issued by public health institutions. On the other hand, with regards to social aspects and user involvement, health-relevant information is not only derived from expert diagnosis, but may also be based on citizens' tweets or symptom reports (see also Boulos et al. 2010; Chew and Eysenbach 2010; Paul and Dredze 2011). These mapping services and platforms for digital disease detection are proposed as supplements to traditional public health surveillance based on biomedical data, not as a substitute. The examined websites as well as comparable efforts in this field indicate some of the promises and challenges of digital public health surveillance. With regards to the initially posed questions—which information is mapped and according to which criteria is it geographically located—the following aspects seem crucial:

Within HealthMap, news items/journalistic reports were predominant among the mapped data. Moreover, these sources were mostly retrieved from commercial news aggregators, and at least for Western countries, Google News items were prevailing. In addition, Chinese language aggregators and sources were indicated as a main source. In light of this observation, it also seems relevant to question the implications of the kind of sources included and excluded in such news aggregators. First of all, this relates to algorithmic selection criteria and hints hence at the gatekeeper function of large-scale, corporate news compilations. Second, not only algorithmic selection criteria, but also governmental censorship could potentially intervene with the reliability of certain sources. In this sense, the selected sources act as gatekeepers, defining in- and exclusion of information. Further research is needed which shows how data sources such as commercial news aggregators act as gatekeepers for information. In particular, it seems relevant to address which gaps and biases might be produced by utilizing such sources.

The choice of sources also indicates issues related to the "digital divide." The lack of mapped information and the blank cartographic spaces for some African countries in HealthMap show that the big data approach depends on criteria that mainly apply to countries for which continuous news reporting and institutional reports are available. In order to function well, it requires a reliable, digital information basis and is hence ideally suitable for societies in which citizens have access to digital technologies and the internet. However, the blank cartographic spaces as such may also be used in order to assess for which countries and with regards to what criteria the big data approach of HealthMap should be expanded.

Similarly, the project FluNearYou that relies on users' symptom reports shows that user involvement is particularly strong in urban regions. It cannot be concretely assessed if this merely reflects population densities in certain regions and higher infection risks in densely populated urban areas, or if residents in rural areas, for instance, are less likely to contribute. While HealthMap mainly relies on journalistic and institutional sources (user submissions are possible, but rare), the North-American community health project has to address several challenges related to users' involvement. First, the website depends on users' willingness to participate in the first place and to do so continuously. Therefore, the project needs to invest in steady efforts to establish, maintain, and expand its community. Its sustainability may be threatened by a mere short-term involvement of users—which may be influenced by an increased interest to report only in the case of illness as well as short-dated interest triggered by media reports. Second, they depend on users' sincerity and capacity to report symptoms. These observations particularly emphasize that user involvement in public health surveillance projects is contingent on digital media literacy as well as basic health education, i.e., the capacity to identify individual symptoms and to report them online.

While the two related projects have been developed and are maintained by independent researchers, the role of corporations seems nevertheless crucial. The observations that both projects receive corporate funding and utilize commercial data sources refers to changing conditions for research: Scholars who aim at implementing practical, digital applications of their research (which is likewise encouraged by governmental funding schemes) often use platforms owned by corporations in order to realize their projects. A main, visually dominant element of FluNearYou and HealthMap is for instance the Google Maps interface. The selected items—based on user reports or digital (commercial) online sources—are located on this corporate, cartographic material (see also Sanjay and Joliveau 2009: 80). Such entanglements between corporate services and independent research also seem problematic, since funding for related research has been provided by Google.org (Freifeld et al. 2010). Therefore, one might question what kind of dependencies may evolve between academic researchers and corporate service providers. Due to the open application programming interface, Google Maps are a popular solution for mapping projects. However, initiatives such as the Ushaidi-Haiti map show that such services are less suitable in the case of certain countries. As Meier recalls in his documentation of volunteers' efforts:

[M]apping this content became more and more challenging because Port-au-Prince was half missing on the *Google Map* of Haiti. The city and roads had not been fully mapped by Google Inc. So some colleagues at OpenStreet-Map crowdsourced the most detailed roadmap of Haiti ever produced in just a matter of days.

(Meier 2012b)

In light of the current focus of FluNearYou and the geographically broad mapping of information in HealthMap that I discussed above, this may not formally be an issue yet. However, further studies concerning the implications of such entanglements between corporate digital services and academic research are needed. This seems particularly relevant since the discussed projects also illustrate the potential benefits and promises that such approaches in digital epidemiological surveillance and public health mapping may offer. While this chapter certainly only opens up the discussion regarding such developments, it aims at highlighting some of the directions in which further technological investments as well as methodological and ethical reflections are needed.

Notes

1 Even earlier, the geographer Charles Picquet created maps for the "Rapport sur la marche et les effets du choléra dans Paris et le département de la Seine" (issued by Louis-François Benoiston de Châteauneuf), which cartographically visualized the number of cholera deaths in Paris during 1832. In this case, however, the maps were not primarily used as tools for investigating eventual causes of the disease (see Châteauneuf 1832: 207ff.).
2 The critical debate concerning neogeography is also reflected in increasingly sceptic discussions on the ideal of participatory culture (see, e.g., Fuchs 2014: 52ff.; Van Dijck 2009).
3 The article is preceded by the following note regarding funding: "The work is funded thanks to a grant from Google.org. The funders had no role in study design, data collection and analysis, decision to publish, or preparation of the manuscript" (Freifeld et al. 2010).
4 See www.frontlinesms.com (accessed July 27, 2017).
5 The website is not accessible anymore, but has been documented at: http://davidrothman.net/2007/02/22/healthmap-epispider (accessed July 27, 2017).
6 An overview of the team's publications can be found on: www.healthmap.org/site/about/publications (accessed July 27, 2017).
7 See also www.healthmap.org/zika/#timeline (accessed July 27, 2017).
8 For an assessment of FluNearYou's accuracy during influenza season, see, e.g., Smolinski et al. (2015).
9 An overview of sources is also provided on the website's "About" section: www.healthmap.org/site/about (accessed July 18, 2016). While Twitter references can be found on the map, the microblogging platform is currently not mentioned as source.
10 Of course, this does not only apply to the quantity, but also the quality of available information.
11 One also needs to keep in mind that less problematic factors, such as population density in rural or urban areas, are relevant here.

12 When searching for alerts by indicating "eyewitness reports" as source, for example, for the max. available time frame of six months (May 25, 2016 until November 25, 2016), no results were available.
13 Despite the condition that users must register with a U.S. or Canadian zip code, on July 20, 2016, a few location marks were placed in Italy, Ukraine, and Russia. Since I did not have access to the user interface, the reasons for this remain unclear.

References

Asokan, G. V. and Asokan, V. (2015) "Leveraging 'big data' to enhance the effectiveness of "one health" in an era of health informatics," *Journal of Epidemiology and Global Health*, 5 (4): 311–314.

Boulos, M. N., Sanfilippo, A. P., Corley, C. D., and Wheeler, S. (2010) "Social web mining and exploitation for serious applications: technosocial predictive analytics and related technologies for public health, environmental and national security surveillance," *Computer Methods and Programs in Biomedicine*, 100 (1): 16–23.

Boyd, D. and Crawford, K. (2012) "Critical questions for big data: provocations for a cultural, technological, and scholarly phenomenon," *Information, Communication & Society*, 15 (5): 662–679.

Brownstein, J. S. and Freifeld, C. C. (2007) "HealthMap: the development of automated real-time internet surveillance for epidemic intelligence," *Euro Surveillance*, 12 (48), available online at: www.eurosurveillance.org/viewarticle.aspx?articleid=3322 (accessed July 27, 2016).

Brownstein, J. S., Freifeld, C. C., and Madoff, L. C. (2009) "Digital disease detection: harnessing the web for public health surveillance," *New England Journal of Medicine*, 360 (21): 2153–2157.

Brownstein, J. S., Freifeld, C. C., Reis, B. Y., and Mandl, K. D. (2008) "Surveillance sans frontières: internet-based emerging infectious disease intelligence and the HealthMap project," *PLoS Med*, 5 (7): e151. https://doi.org/10.1371/journal.pmed.0050151.

Burgess, J. E. and Green, J. B. (2009) "The entrepreneurial vlogger: participatory culture beyond the professional-amateur divide" in Snickars, P. and Vonderau, P. (eds) *The YouTube reader*, Stockholm: National Library of Sweden, 89–107.

Centers for disease control and prevention (2014) "The flu season," available online at: www.cdc.gov/flu/about/season/flu-season.htm (accessed July 27, 2016).

Châteauneuf, B. de (1832): *Rapport sur la marche et les effets du choléra dans Paris et le département de la Seine*, Paris: Impr. Royale, available online at: http://gallica.bnf.fr/ark:/12148/bpt6k842918/f3.item (accessed July 27, 2017).

Chew, C. and Eysenbach, G. (2010) "Pandemics in the age of Twitter: content analysis of tweets during the 2009 H1N1 outbreak," *PLoS one*, 5 (11): e14118. https://doi.org/10.1371/journal.pone.0014118.

Chunara, R., Aman, S., Smolinski, M., and Brownstein, J. S. (2013) "Flu Near You: an online self-reported influenza surveillance system in the USA," *Online Journal of Public Health Informatics*, 5 (1): e133, available online at: http://europepmc.org/articles/pmc3692780 (accessed July 27, 2016).

Chunara, R., Andrews, J. R., and Brownstein, J. S. (2012) "Social and news media enable estimation of epidemiological patterns early in the 2010 Haitian cholera outbreak," *The American Journal of Tropical Medicine and Hygiene*, 86 (1): 39–45.

Clarke, G. P. and Langley, R. (1996) "A review of the potential of GIS and spatial modelling for planning in the new education market," *Environment and Planning C: Government and Policy*, 14 (3): 301–323.

Collier, N., Doan, S., Kawazoe, A., Goodwin, R. M., Conway, M., Tateno, Y., Ngo, Q. H., Dien, D., Kawtrakul, A., Takeuchi, K., Shigematsu, M., and Taniquchi, K. (2008) "BioCaster: detecting public health rumors with a web-based text mining system," *Bioinformatics*, 24 (24): 2940–2941.

Dalton, C., Taylor, L. and Thatcher, J. (2016) "Critical data studies: a dialog on data and space," *Big Data and Society*, 3 (1): 1–9. https://doi.org/10.1177/2053951716648346.

Dalton C. and Thatcher, J. (2014) "What does a critical data studies look like and why do we care? Seven points for a critical approach to 'big data'," available online at: http://societyandspace.com/2014/05/19/dalton-and-thatcher-commentary-what-does-a-critical-data-studies-look-like-and-why-do-we-care (accessed July 27, 2016).

Deuze, M. (2006) "Ethnic media, community media and participatory culture," *Journalism*, 7 (3): 262–280.

Eke, P. I. (2011) "Using social media for research and public health surveillance," *Journal of Dental Research*, 90 (9), 1045–1046.

Eysenbach, G. (2009) "Infodemiology and infoveillance: framework for an emerging set of public health informatics methods to analyze search, communication and publication behavior on the internet," *Journal of Medical Internet Research*, 11 (1): e11. https://doi.org/10.2196/jmir.1157.

Freifeld, C. C., Chunara, R., Mekaru, S. R., Chan, E. H., Kass-Hout, T., Iacucci, A. A., and Brownstein, J. S. (2010) "Participatory epidemiology: use of mobile phones for community-based health reporting," *PLoS Med*, 7 (12): e1000376. https://doi.org/10.1371/journal.pmed.1000376.

Freifeld, C. C., Mandl, K. D., Reis, B. Y., and Brownstein, J. S. (2008) "HealthMap: global infectious disease monitoring through automated classification and visualization of internet media reports," *Journal of the American Medical Informatics Association*, 15 (2): 150–157.

Fuchs, C. (2014) *Social media: a critical introduction*, Los Angeles: Sage.

Ginsberg, J., Mohebbi, M. H., Patel, R. S., Brammer, L., Smolinski, M. S., and Brilliant, L. (2008) "Detecting influenza epidemics using search engine query data," *Nature*, 457: 1012–1014. https://doi.org/10.1038/nature07634.

Goodchild, M. (2009) "NeoGeography and the nature of geographic expertise," *Journal of location based services*, 3 (2): 82–96.

Haklay, M. M. (2013) "Neogeography and the delusion of democratisation," *Environment and Planning A*, 45 (1): 55–69.

Hay, S. I., George, D. B., Moyes, C. L., and Brownstein, J. S. (2013) "Big data opportunities for global infectious disease surveillance," *PLoS Med*, 10 (4): e1001413. https://doi.org/10.1371/journal.pmed.1001413.

HealthMap Brochure (n.d.) "Global health, local knowledge," available online at: www.healthmap.org/print_materials/brochure.pdf (accessed July 27, 2016).

Hermida, A. (2012) "Tweets and truth: journalism as a discipline of collaborative verification," *Journalism Practice*, 6 (5): 659–668.

Jenkins, H. (2006) *Fans, bloggers, and gamers: Exploring participatory culture*, New York: New York University Press.

Jordan, T. (2015) *Information politics: liberation and exploitation in the digital society*, London: Pluto Press.

Jost, C. C., Mariner, J. C., Roeder, P. L., Sawitri, E., and Macgregor-Skinner, G. J. (2007) "Participatory epidemiology in disease surveillance and research," *Revue Scientifique Et Technique-Office International Des Epizooties*, 26 (3): 537–547.

Kay, M., Santos, J., and Takane, M. (2011) "mHealth: new horizons for health through mobile technologies," available online at: www.who.int/goe/publications/goe_mhealth_web.pdf (accessed July 27, 2017).

Keller, M., Blench, M., Tolentine, H., Freifeld, C. C., Mandl, K. D., Mawudeko, A., Eysenbach, G., and Brownstein, J. S. (2009) "Use of unstructured event-based reports for global infectious disease surveillance," *Emerging Infectious Diseases*, 15 (5): 689–695.

Kitchin, R. (2014a) *The data revolution: big data, open data, data infrastructures and their consequences*, London: Sage.

Kitchin, R. (2014b) "Big data, new epistemologies and paradigm shifts," *Big Data & Society*, 1 (1): 1–12. https://doi.org/10.1177/2053951714528481.

Kotz, D. (2015) "Digital health map tracks new Ebola cases in real time," available online at: www.bostonglobe.com/lifestyle/health-wellness/2014/08/07/digital-health-map-tracks-new-cases-ebola-virus-real-time/DjPXtj8O7R81wij7dhDkIN/story.html (accessed July 27, 2016).

Lazer, D. and Kennedy, R. (2015) "What can we learn from the epic failure of Google Flu Trends?," available online at: www.wired.com/2015/10/can-learn-epic-failure-google-flu-trends/ (accessed July 27, 2016).

Lazer, D., Kennedy, R., King, G., and Vespignani, A. (2014) "The parable of Google Flu: traps in big data analysis," *Science*, 343 (6176): 1203–1205.

Lupton, D. (2013) "Quantifying the body: monitoring and measuring health in the age of mHealth technologies," *Critical Public Health*, 23 (4): 393–403.

Lupton, D. (2014) *Digital sociology*, London: Routledge.

Lupton, D. (2015) "The thirteen Ps of big data," available online at: https://simplysociology.wordpress.com/2015/05/11/the-thirteen-ps-of-big-data/ (accessed July 25, 2017).

Lyon, A., Nunn, M., Grossel, G., and Burgman, M. (2012) "Comparison of web-based biosecurity intelligence systems: BioCaster, EpiSPIDER and HealthMap," *Transboundary and Emerging Diseases*, 59 (3): 223–232.

Marz, N. and Warren, J. (2012) *Big data: principles and best practices of scalable real-time data systems*, Westhampton: Manning.

McLeod, K. S. (2000) "Our sense of Snow: the myth of John Snow in medical geography," *Social Science & Medicine*, 50 (7–8): 923–935.

Meier, P. (2012a) "Crisis mapping in action: how open source software and global volunteer networks are changing the world, one map at a time," *Journal of Map & Geography Libraries*, 8 (2): 89–100.

Meier, P. (2012b) "How crisis mapping saved lives in Haiti," *National Geographic*, available online at: http://voices.nationalgeographic.com/2012/07/02/crisis-mapping-haiti (accessed July 27, 2016).

Meier, P. (2014) "Crisis mapping Haiti: some final reflections," available online at: http://reliefweb.int/report/haiti/crisis-mapping-haiti-some-final-reflections (accessed July 27, 2016).

Mittelstadt, B. D. and Floridi, L. (2016) "The ethics of big data: current and foreseeable issues in biomedical contexts," *Science and Engineering Ethics*, 22 (2): 303–341.

Morrow, N., Mock, N., Papendieck, A., and Kocmich, N. (2011) "Independent evaluation of the Ushahidi Haiti project," available online at: www.alnap.org/resource/6000 (accessed July 27, 2016).

Mosco, V. (2015) *To the cloud: big data in a turbulent world*, London: Routledge.

Newsom, S. W. (2006) "Pioneers in infection control: John Snow, Henry Whitehead, the Broad Street pump, and the beginnings of geographical epidemiology," *Journal of Hospital Infection*, 64 (3): 210–216.

Norris, P. (2001) *Digital divide: civic engagement, information poverty, and the internet worldwide*, Cambridge: Cambridge University Press.

O'Reilly, T. (2007) "What is Web 2.0: design patterns and business models for the next generation of software," available online at: http://papers.ssrn.com/sol3/Papers. cfm?abstract_id=1008839 (accessed October 16, 2016).

Ostfeld, R. S., Glass, G. E., and Keesing, F. (2005) "Spatial epidemiology: an emerging (or re-emerging) discipline," *Trends in Ecology & Evolution*, 20 (6): 328–336.

Paneth, N. (2004) "Assessing the contributions of John Snow to epidemiology: 150 years after removal of the Broad Street pump handle," *Epidemiology*, 15 (5): 514–516.

Paul, M. J. and Dredze, M. (2011) "You are what you tweet: analyzing Twitter for public health," *ICWSM*, 20: 265–272.

Petersen, S. M. (2008) "Loser generated content: from participation to exploitation," available online at: http://firstmonday.org/article/view/2141/1948 (accessed July 27, 2016).

Polgreen, P. M., Chen, Y., Pennock, D. M., Nelson, F. D., and Weinstein, R. A. (2008) "Using internet searches for influenza surveillance," *Clinical Infectious Diseases*, 47 (11): 1443–1448.

Rao, L. (2013) "Why we need to kill 'big data'," available online at: https://techcrunch. com/2013/01/05/why-we-need-to-kill-big-data (accessed July 27, 2016).

Richterich, A. (2016) "Using transactional big data for epidemiological surveillance: Google Flu Trends and ethical implications of 'infodemiology'" in Mittelstadt, B. D. and Floridi, L. (eds) *Ethics of biomedical big data*, London: Springer, 41–72.

Sanjay, R. and Joliveau, T. (2009) "NeoGeography: an extension of mainstream geography for everyone made by everyone?," *Journal of Location Based Services*, 3 (2): 75–81.

Science and Technology Committee (2015) "The big data dilemma," available online at: https://publications.parliament.uk/pa/cm201516/cmselect/cmsctech/468/468.pdf (accessed July 27, 2016).

Shapiro, I., Brin, C., Bédard-Brûle, I., and Mychajlowycz, K. (2013) "Verification as a strategic ritual: how journalists retrospectively describe processes for ensuring accuracy," *Journalism Practice*, 7 (6): 657–673.

Signorini, A., Segre, A. M., and Polgreen, P. M. (2011) "The use of Twitter to track levels of disease activity and public concern in the US during the influenza A H1N1 pandemic," *PLoS One*, 6 (5): e19467. https://doi.org/10.1371/journal.pone.0019467.

Smolinski, M. S., Crawley, A. W., Baltrusaitis, K., Chunara, R., Olsen, J. M., Wójcik, O., Santillana, M., Nguyen, A., and Brownstein, J. S. (2015) "Flu Near You: crowdsourced symptom reporting spanning 2 influenza seasons," *American Journal of Public Health*, 105 (10): 2124–2130.

Soden, R. and L. Palen (2014) "From crowdsourced mapping to community mapping: the post-earthquake work of OpenStreetMap Haiti" in Rossitto, C., Ciolfi, L., Martin, D. and Conein, B. (eds) *COOP 2014: proceedings of the 11th international conference on the design of cooperative systems, 27–30 May 2014, Nice (France)*, London: Springer, 311–326.

Stoové, M. A. and Pedrana, A. E. (2014) "Making the most of a brave new world: opportunities and considerations for using Twitter as a public health monitoring tool," *Preventive Medicine*, 63: 109–111. https://doi.org/10.1016/j.ypmed.2014.03.008.

Sui, D., Goodchild, M., and Elwood, S. A. (2013) "Volunteered geographic information, the exaflood, and the growing digital divide" in Sui, D., Goodchild, M., and Elwood, S. (eds) *Crowdsourcing geographic knowledge*, Amsterdam: Springer Netherlands, 1–12.

Tene, O. and Polonetsky, J. (2012) "Big data for all: privacy and user control in the age of analytics," *Northwestern Journal of Technology and Intellectual Property*, 11 (5): 249–273.

Tolentino, H., Kamadjeu, R., Fontelo, P., and Liu, F. (2007) "Scanning the emerging infectious diseases horizon: visualizing ProMED emails using EpiSPIDER," *Advances in Disease Surveillance*, 2: 169, available online at: https://lhncbc.nlm.nih.gov/files/archive/pub2007055.pdf (accessed July 27, 2016).

Turner, A. (2006) *Introduction to neogeography*, New York: O'Reilly Media.

Van Dijck, J. (2009) "Users like you? Theorizing agency in user-generated content," *Media, Culture & Society*, 31 (1): 41–58.

Vayena, E., Salathé, M., Madoff, L. C., and Brownstein, J. S. (2015) "Ethical challenges of big data in public health," *PLoS Comput Biol*, 11 (2): e1003904. https://doi.org/10.1371/journal.pcbi.1003904.

Velasco, E., Aqheneza, T., Denecke, K., Kirchner, G., and Eckmanns, T. (2014) "Social media and internet-based data in global systems for public health surveillance: a systematic review," *Milbank Quarterly*, 92 (1): 7–33.

Warschauer, M. (2004) *Technology and social inclusion: rethinking the digital divide*, Cambridge, MA: MIT Press.

Weaver, D. A., and Bimber, B. (2008) "Finding news stories: a comparison of searches using LexisNexis and Google News," *Journalism & Mass Communication Quarterly*, 85 (3): 515–530.

Wójcik, O. P., Brownstein, J. S., Chunara, R. and Johansson, M. A. (2014) "Public health for the people: participatory infectious disease surveillance in the digital age," *Emerging Themes in Epidemiology*, 11 (1): 1–7. https://doi.org/10.1186/1742-7622-11-7.

Zhang, Y., Dang, Y., Chen, H., Thurmond, M., and Larson, C. (2009) "Automatic online news monitoring and classification for syndromic surveillance," *Decision Support Systems*, 47 (4): 508–517.

Ziemke, J. (2012) "Crisis mapping: the construction of a new interdisciplinary field?," *Journal of Map & Geography Libraries*, 8 (2): 101–117.

Zikopoulos P., Eaton, C., DeRoos, D., Deutsch, D., and Lapis, G. (2012) *Understanding big data*, New York: McGraw Hill.

Index

Page numbers in *italics* denote tables.